Surf, Flood, Fire & Mud

How the strongest El Niño ever
led to years of extremes,
and what this means going forward.

By: Nathan Cool

KDP ISBN: 9781980295303
REV180313

Cover image courtesy of NASA, taken February 20, 2017 showing one of the atmospheric rivers that led to flooding in California.

To all of those who suffered from the
ravaging floods, wildfires, and mudslides
in California and around the world,
and to the memories of those
who perished, as well as those
who risked their lives to save others.

May you never be forgotten.

Table of Contents

Prologue 9

Portents 15

Backstory 27
- When Waters Warm 29
- The Eye of the Storm 32
- Explosive Forces 35
- Four Years of Niña's Reign 36

Surf 39
- The Biggest Wave 43
- Niño y Nazaré 45
- Titans and Bellwethers 48
- SHARK! 51
- Snakes on the Strand 54
- Crabs 56
- Slugs 57
- The Dead Zones 58
- What Comes Up 61
- The Cyclone Signal 64

Flood 69
- The PDO 75
- The Blob 76
- History's Lesson 79
- When Atmospheric Rivers Flow 81
- The MJO 85
- Oroville by Example 89
- Harvey's Alarms 94
- The Irma Conflict 101
- The Maria Test 105
- Between the Cracks 110
- The Calm Before 114

Fire 117
- Yesteryear 120

Fanning the Flames..125
The Niño Connection.....................................128
The Surf Signal...131
California Under Siege 135
The Thomas Fire..139
When the World Burned...............................148

Mud...153
Montecito..156
The Near Miss...165
Big Sur..168
La Conchita...170
Lessons from Sierra Leone 174

Communication Breakdown 177

Epilogue 193

Weather Charts...207
Monster Jaws 209
East Coast Slammer......................................211
Three Atlantic Hurricanes.............................213
The Wind Machine 215
Montecito Slide...217

Sources..219

Notes & Tidbits...221

Surf
Flood
Fire
Mud
Salo
Inundo
Ignis
Lato

Prologue

Look deep into nature, and then,
you will understand everything better.

~Albert Einstein

The story you are about to read is true, concerning extreme events, horrific tragedies, and fatal mistakes that started in 2015 and continued for the next few years. Though sometimes shocking and unimaginable, nothing has been fabricated or exaggerated. Accounts of these events are often disturbing, but all are factual, not falsified. Yet while many of the events detailed in this book resulted in death and destruction over the last few years, memories of them are already starting to fade from many who lived through the worst El Niño ever, and the years of disasters that followed.

Time and memory are funny things. As one develops, the other dissolves. It may be difficult to immediately remember the events on December 11, 2015, June 17, 2017, August 17, 2017, December 4, 2017, or perhaps January 8, 2018. Yet all of those dates signify record-breaking (or near record-breaking) events from destructive surf in California, deadly floods in Texas, ravaging wildfires in California, and cataclysmic mudslides in Montecito, California. It's completely natural to have less than total recall on events like these until we're reminded of them. Human nature has a way of biasing our cognition to single-out catastrophes that impact us directly, while fading to forgetfulness those with less proximity to our daily lives. Looking back though can help to form our future, and keep us safe.

Besides the sometimes-faulty memory inherent to our human fallibility, it can be difficult as well to care about climate around the world when we barely remember weather in the tiny corner of our own. Weather-driven events from 2015 into 2018 left traumatic scars for many, but these too will no doubt fade over time as it becomes harder to recall the 20-foot-plus waves that pounded, flooded, and damaged the coast of California in December 2015; the devastation left by Hurricanes Harvey, Irma and Maria in 2017; the Thomas fire in December 2017 preceded by 250 simultaneous Northern California wildfires in October of that year; or unprecedented rain rates that caused the fatal mudslides in Montecito. Strangely enough, all but the first (high surf) occurred *after* a record breaking El Niño had quietly crept into the record books, and things were — statistically speaking — supposed to be quieting down. But they didn't. The reasons why may surprise you.

Stories have a way of either being stretched or sadly forgotten — even those from just a few years, or months, ago. Being employed to watch weather across the Pacific, my reporting over the past couple of decades has seen and told its fair share of impressive episodes of nature's wrath. Yet none compare to what happened over the last few years, when El Niño returned with a vengeance, kicking off a chain of disastrous events that continue to this day. It's a story worth telling and remembering, knowing this will happen again.

While this book tells stories of weather and climate cycles that caused recent extremes and subsequent disasters, I'd like to clarify something. To clear the air, let me start by saying that climate change is real and we know that humans play a large part in it. However, one needs to be cautious contributing *all* weather events (and tragedies that follow) to human-induced global warming. While human-induced climate change is a reality and needs to be mitigated, nature plays a strong role in controlling what happens as well, sometimes with year-to-year variations, but also from decades- to centuries-long cycles that, occurring so infrequently, could be mistaken for something else. Throughout this book I'll touch on the issue of climate change from time to time, but only when needed to sort out what can and what cannot be directly attributed to human-induced global warming. This is not meant as fodder for climate deniers; instead, it is a way to better understand what caused recent catastrophes, allowing us to then better learn how to manage them. If one blames everything on climate change, then a serious problem quickly becomes akin to a boy crying wolf, falling on deaf ears that should otherwise listen closely to the sounds of a warming world — and other issues we should prepare for. If all we did was concentrate our efforts on reducing our carbon footprints, then we'd be caught with our pants down when natural cycles and events get underway. We need to be aware of both a changing climate and the inherent wrath of nature. We need to plan for both, and mitigate the risks accordingly.

As I tell the tale of how nature ran a course of cataclysmic wrath starting in 2015, the human element is interwoven time and again — not always as a victim, but sometimes perpetrator. We indeed have a connection to nature, but it's more than an ethereal sense of wonder or oneness; it's also a symbiotic relation that wraps our human tendrils into places they shouldn't be — with eyes blind and ears deaf to what the world tries to tell us. In too many cases humans are to blame for mishaps that made recent tragedies worse than they had to be. Fatal human errors actually caused some disasters, and mishaps unfolded as well by those in charge of keeping us safe. Humans play a major role in what happened over the past few years, in more ways than one. Knowing what went wrong can help us sort out the things that went right, giving us further insight into how we can better deal with not just nature, but our own issues in the future as well.

To start things off, I'll take a look at harbingers that signaled the impending spat of disasters that, although starting in 2015, were ringing alarm bells long before then. The "Portents" chapter will get the ball rolling, but then the "Backstory" chapter will give a quick overview of how El Niño played a role, with a smidge of science thrown in to understand why this climate cycle does what it does. This sets the stage for the main event, consisting of four acts of nature.

The next four chapters will cover both the stories and science of not just *what* happened, during the last few years of natural disasters, but also the *when*, *where*, and *why*. Each chapter, *Surf*, *Flood*, *Fire*, and *Mud*, will recount the events that occurred and led up to such disastrous events, how this compared to years past, and various studies that help us understand the presence and/or absence of causal links.

After covering the four elements comprising El Niño's chain of events and disasters, I'll take a look at why many were caught unaware. Alarms were sounded, but they were silent to many for a number of reasons. Originally federally-funded programs failed, one of which I worked on alongside NOAA and National Weather Service staff a few years ago. I'll discuss my personal experience working with NOAA and the NWS, and explain what went wrong. But that part of the story has a silver lining: Today there are new, better systems in place, yet many don't realize they can utilize them. Nevertheless, in some cases, trusted officials dropped the ball on keeping us informed, despite the advent of newer alert systems. I'll get to all of that in the "Communication Breakdown" chapter.

I'll then wrap things up with a quick epilogue — a chance for me to rant while looking at how the never-ending story of El Niño, climate, and subsequent weather events require a certain type of attention, and action. This epilogue is not my Kaczynskian, backwoods, tinfoil hat-wearing manifesto, but I admittedly change my tone in this chapter as I share my personal take on the politicization and polarization of climate and weather, and why deniers and alarmists would behoove themselves (and others) by stepping back into their respective corners, and take a few deep breaths.

To let a picture say a thousand words (in a weather kind of way), I've also included a few weather charts in the back-matter section of this book. I used classic, old-school style charts from NOAA, which at first glance may look benign and boring. But there is a chilling contrast to these simple, black-and-white drawings when you realize the magnitude of the events they represent.

Along the way, you'll see numerous superscript numbers referring to endnotes, which you'll find in the "Notes & Tidbits" section near the end of this book. Given that much of the information in this book could be shocking and perhaps inconceivable, I wanted to include the exact reference points throughout the book so that, if you so desire, you can fact-check these items yourself. I've also included some interesting tidbits in this section of references including how some things were calculated, word origins, and more.

Yet the story you're about read is not meant to stir fear, purport worst-case scenarios, or point fingers of blame. This isn't a story about politics, conspiracies, or apocalyptic doom. This isn't a story about what's wrong in the world, or endless rants of how things could, should, or perhaps be.

Instead, this is a story about recently wild weather — what caused it, led up to it, unfolded, and was left in its wake. It's a story of science, human fallibility, fatal mistakes, and the will to push forward in the face of adversity. It's a story about you, me, and everyone on the planet, inextricably connected by forces of nature that, while not in our control, can be dealt with. This is a story of four kinds of hell that touched our world over the past few years, kicked off by a recent, record-breaking El Niño. This is a story of not just what happened, but also what followed. This is the story of *Surf, Flood, Fire & Mud.*

Portents

We must free ourselves of the hope that the sea will ever rest. We must learn to sail in high winds

~Aristotle Onassis

The warning signs came early. Yet for many, they were soon forgotten. Nevertheless, the harbinger of change became evident two decades ago on January 30, 1998, which started an epoch of weather awareness — but not necessarily preparedness. This new era's kickoff in the late '90s was recently punctuated by three strange years — from 2015 into 2018 — when California's weather and surf got turned on its head. But January '98 was when it all began; it's the day that put El Niño on the map, ringing an alarm that went seemingly silent for seventeen years after. While some Californians had heard of El Niño back in the day, many had no idea what this climate phenomenon was, or was capable of unleashing. That all changed rather quickly, and the future would never quite be the same. January 30, 1998 was a milestone; yet it was just the beginning.

On that fateful Friday in '98 — an otherwise unassuming day for weather in Southern California — energy from a gigantic Pacific storm, which had formed thousands of miles away a week earlier, was about to rock the California coastline with massive waves, bringing surf, and destruction, that would soon become historic. There was an ominous dark cloud hovering over a silver lining of surf-able hope. Monstrous waves were coming, many would ride them, but there would be sad aftermath.

Marked in surf history as "Big Wednesday", Hawaii first saw the impacts from this behemoth Pacific storm two days before California did, making the now-infamous break off Maui's north shore known as "Jaws" a legend. *National Geographic* magazine put a now-memorable shot of big wave surfer Laird Hamilton riding a monstrous Jaws bomb on its November 1998 cover. The biggest waves though weren't ridden at Jaws on Big Wednesday; that credit goes to Ken Bradshaw who rode the biggest wave ever surfed up to that point, clocking in at approximately 70 feet at Outer Log Cabins — a tow-in surfing break off the north shore of Oahu known for its gargantuan waves.[1] That record would be broken when the El Niño of 2015-16 stirred chaos in the Pacific and challengers took on colossal waves in early 2016. But back in January '98, Laird, along with Buzzy Kerbox, Dave Kalama and others, strapped onto high-performance surfboards, and with the assistance of jet skis were towed into 50-foot slabs of Pacific surf at Jaws, eventually dropping their tow-in lines to ride, in solo, some of the biggest waves in the world.

The January 1998, Big Wednesday swell hitting Hawaii was a warning for the west coast of the United States. California was next, and about to experience a day so memorable that it earned the SoCal title "Big Friday" — a day that now serves as baseline from which most major California swells are compared. But surf wasn't the only concern that day. It was the start of something big — literally.

While strong waves had reached the California coast earlier during the last week of January 1998, it was on Friday the 30th that seas off Point Conception, near Santa Barbara, reached 16 feet, carrying such intense, long-period energy focused on Southern California that breaking waves had the potential to nearly double in size before breaking near the shore. By early morning, beaches along the California coast were evacuated as gargantuan breakers crashed with such force that their surge washed frothy remnants far up onto California beaches. By noon, the California Highway Patrol closed sections of the PCH as rocks and coastal cobble were tossed onto California's iconic coastal roadway — a result of energy released with such intensity that it shook the ground a great distance from the water's edge. By midafternoon, many of the oldest and weakest structures lining the historic coastline of Southern California could take no more; the damage had begun.

The Ventura Pier, which, at the time, jutted out nearly 2000 feet from land to the deep blue, was quickly curtailed to nearly two-thirds its size as incessant, fast moving, 20-foot-plus waves smashed its wooden pilings for hours on end. Coastal waters throughout California were littered with debris, not from rain-driven runoff (it was dry in SoCal that day), but from outflow of waterfront waste reclaimed by the advancing surf, devouring terra firma's jetsam back out to sea. Remnants of the Ventura Pier added to this collection of hazardous water debris. But this swell, fueled by El Niño, was only a portent of things to come that would make El Niño not just an indelible term in the weather forecasting lexicon, but a phrase to be feared with wide-eyed respect when uttered by climate scientists and meteorologists for many years to come.

Shortly after Big Friday, the month of February 1998 brought historic, flooding rains to California. Mid-winter storms, heavily laden with perceptible water, were so intense that rivers were widened from a deluge of full-grown trees, ripped from the banks where they were once firmly rooted. Myriad refuse and rubble that lay dormant for decades were being flushed to the sea from an unstoppable force, taking along with it everything that lie, or fell, in its path. So high were piles of driftwood on Southern California beaches from tides eventually pushing storm destruction back to land, that it took special cleanup efforts more than a year to dispose of all of nature's diversified, storm-driven detritus. All of this was attributed to a climate pattern with a Spanish name that everyone around the world — and especially in California — from this point forward, would translate with ease but state with trepidation: *El Niño*.

The events of 1998 burned memories forever in the mind of most Californians, who sat on the edge of their seats with nervous anticipation whenever the words El Niño were spoken. As anxiety grew over the next seventeen years, so did concern that much-needed rain for desert areas comprising the American Southwest could once again cause disasters racking-up costs exceeding the epic $4.2 billion from the 1997-98 event. So it was no surprise that when a NASA scientist, during a media interview in 2015 called that year's growing El Niño a "Godzilla", news media lit up with weather shock-and-awe. Signs for the next round of ominous Pacific wrath were watched with bated breath. But the wait would be long, and many would turn their backs on a proverbial sleeping giant.

Preparations for a cataclysmic El Niño were well underway by the end of 2015, especially since there was a new, potentially sinister snag in the Pacific forecast given another fictive epithet, the "Blob" — a bizarre warmth in the North Pacific (which I'll talk about more in the "Flood" chapter). Combined with a known, decades-long shift in the Pacific that could accentuate El Niño (known as the PDO, which I'll get to later as well), it appeared that a perfect-storm kind of El Niño scenario was inevitably going to ravage the west coast of the United States. With warranted reverence to a warming world where weather history could repeat itself with astronomical consequences, a sky-is-falling hysteria overtook much of the west coast where devastating floods could quickly ruin a way of life — as they have in the past. But as the weather-weary waited anxiously in 2015, anticipation turned not to terror, but instead quiet complacency as only a few storms brought any rain to the desiccated regions of California and the desert southwest. The "Godzilla" El Niño during the winter of 2015-16, hyped by media and so many otherwise cautious forecasters and climatologists, became a paper-tiger that merely yawned. Godzilla slept — or so it seemed.

All the while, a smaller, west coast cadre of ocean weather watchers knew very well that El Niño had arrived in 2015 — not with a whimper, but a very loud bang. Surfers up and down the coast of California, Oregon, and Washington, experienced surf that well-surpassed the events of 1998. Hawaii also experienced surf unlike *normal* winters, but for the remaining, sans-surf population, the El Niño of 2015-16 looked like a bust — a little boy crying wolf, a forecast blunder, and basically an overhyped weather dud. But no matter if you rode boards or merely longed for drought-relief, nearly everyone asked the same question: *Is this the new normal – is this the new Niño?*

For more than 20 years I've tracked every storm across the Pacific that would affect surf in California, Oregon, Washington, Hawaii, and Central America. Starting this profession back in 1996, watching gales from Antarctica to Alaska and Japan to California, I witnessed in 1997-98 what some said would be just another hum-drum year for surf and weather — it wasn't. Not fully believing what I was seeing in January '98 from then-nascent, internet accessible weather models, I reported quite cautiously that epic surf was due in the next seven to ten days from intense, Western Pacific-originating storms. I calculated one momentous event with an ETA for January 30, 1998 in Southern California, calling for surf that no one would soon forget. Many thought I was off my freakin' rocker; perhaps *I* was the one crying wolf, or hyping my forecasts for the benefit of higher ratings. Despite beratement from readers, I persisted to dot my i's and cross my t's, double checking and triple checking my forecast calculations. Yet no matter how many times I ran the numbers, no matter how many ways I calculated the swell from various wave and weather models, nothing changed. The math remained the same. Strong storms were headed across the Pacific in January '98, and it may not be pretty.

My job as a forecaster is often rewarded with belated thanks. Only when something *does* happen, I'll get some kudos...if I got it right. Thankfully, I was eventually lauded for sticking to the facts and not faltering in the face of criticism. Big Wednesday (and Friday) was also a turning point, where I vowed never to take my forecasting job with a grain of salt. El Niño was a serious issue that deserves serious reporting — not hype, but facts. Despite epic surf that gives wave riders the widest of smiles and eyes, lives are lost, properties are destroyed, and businesses are harmed with financial burden. Unlike forecasting to help decide the best day to call off work and catch some surf-able stoke, El Niño forecasting, I felt, was a far more serious issue to deal with.

So you might imagine that when I got wind of the Godzilla and Blob hype permeating the media in 2015, I cringed, shook my head, and grit my teeth, especially when seeing cartoons and internet memes showing mythical monsters rising from the ocean depths with comical catch-phrases. In 2015 — seventeen years after Big Friday and the stormy destruction that followed — El Niño was becoming a point of ridicule and humor. It was disheartening, and disappointing. But instead of getting hot under the collar I put my nose ever closer to the proverbial grindstone to once again monitor what could be a historic event. Yet as I did, and the El Niño of 2015 started to unfold, things just weren't adding up.

The El Niño of 2015-16 would indeed be different, but in ways no one could quite expect. This time around we were in for surprises that would linger into the following winter of 2016-17 when record rain fell, leading to a domino effect in 2017-18 of wild wind-driven fires feeding on the then-lush brush, later wiped out when the rains returned, bringing deadly torrents of mud and boulders in January 2018. El Niño didn't just *happen*; it created a multi-year ripple effect signaling a slippery slope of a changing climate.

Sadly, there has been a *crying wolf* mentality since the major Niño of 1997-98 that left a major majority immune to catastrophic cries of climate change, and El Niño. Hyperbole replaced headlines over the years, watering down the seriousness of El Niño. To get ahead of the hype and provide my readers with facts on what El Niño was (and could possibly evolve into), I studied and monitored this weather phenomenon with greater intensity than ever before, and for good reason. Following Big Friday of '98 I received yet another barrage of emails from my readers — this time with less beratement — wondering if climate change (which was barely on the minds of many back in 1998) could indeed be real, and possible impetus for the strong El Niño that year. To question climate change as a factor today is almost unheard of, but just a couple decades ago it was still somewhat uncharted territory.

Curiosity of Earth's climate in the time of Big Friday was justified, as — although hard to imagine now — global warming (as it was more popularly called back then) was seen by many laypersons back in the '90s as a theory with little proven fact. That, though, was an anecdotal misconception, primarily perpetrated by the same media that would later flip-flop to doomsday scenarios a couple decades later. Ice loss was minimal in the '90s compared to today; sea levels hadn't risen or expanded to the extent that have now; and droughts, fires, and floods were still seen as typical, *expected* events. In fact, the mid 1990s was preceded by a time when media disputed the possibility of a warming world. A generation prior to Big Friday and the 1997-98 Niño saw cover stories from *Time* and *Newsweek* magazines in the 1970s that raised contradicting concern. Instead of a warming world, these magazines were purporting the idea of consensus for just the opposite: an impending ice age.[2] *Newsweek* in 1974 reported that, because of an upcoming ice age, we could see a drastic decline in food production with serious political implications, and *Time* reported that the world's atmosphere was continually cooling — not warming. This mindset crossed over into the next generation, as those 1970s' headlines, from the top names in news, sent fears of cold weather and high-cost winter heating across America's then climate-conscious citizens. In the 1970s, many were more concerned with a cooling- not warming-world. But this was truly (pardon the expression) "fake news". The media obviously had it wrong. We weren't cooling; we were getting warmer. We've come a long way since then.

Big Friday and the El Niño of 1997-98 occurred in a different era of weather knowledge, history, greenhouse gas accumulations, melting ice caps, and sea level rise. This was also just a few years following the initial (and not widely publicized) first assessment report by the fairly new Intergovernmental Panel on Climate Change (the IPCC). Three years later, in 2001 — twenty-seven years after *Newsweek* said we were doomed from a cooling planet — the IPCC would issue their third assessment report, which would be the basis for Al Gore's *An Inconvenient Truth*; his magnum opus on climate doom. It's no wonder that many people following the events of 1997-98 had more than one question concerning our climate that deserved an *answer*, not something that they saw as a *theory*. Luckily, science was establishing irrefutable proof, aligning with Gore in many regards.

But time, and our perceptions of it, would lead to many turning a blind eye. And over time, science would change.

With climate change and El Niño taking center stage as the new millennium commenced, I published a lighthearted (yet serious) thesis in my book *Is it Hot in Here: The simple truth about global warming*. Releasing this book in 2006 wasn't the best decision I ever made as a writer, as I was jumping on a soon to be crowded bandwagon, filled with polarized pundits swaying either to alarm or denial, with no room for anything in-between. Back then, an ex-VP was attracting a much wider audience, which, although not the best thing for *my* book sales, did bring awareness to a warming world. Yet Al Gore, while bringing attention to climate change and starting the wheels of believability rolling, was lambasted (justifiably) more than once by scientists on the over-exploitation of extreme worst-case scenarios, cherry picked from the IPCC's third assessment report. Stating the IPCC extremes would be normals and not the outliers that science had specified, Gore was criticized for over-pegging the hype-meter — and he defiantly rationalized doing so.[3] His momentum wouldn't waiver. Gore was on a mission, and the world was listening — for a time.

Polar bears soon became the new millennium's oppressed poster children for climate change.[4] People began trading in Hummers for carbon-friendly Priuses. The solar panel business would soon be booming, and people, overall, were becoming more aware of their *carbon footprint* — a nascent term in ecological word-stock that is now a very P.C. thing to discuss (as is the term P.C., making P.C. a redundantly recursive subject at dinner parties and the basis for an entire season of *South Park*).[5] It was becoming *en vogue* to be climate conscious, telling tales of your ecological efforts as a way to wear your socially-pressured badge of climate honor.

As the new millennium got well underway, the climate mindset had changed; when anything had even the slightest hint, echo, or smell of being a result of humankind's climate negligence, people took notice. I did too.

As Niños lightly waxed and waned over the next couple of decades following the 1997-98 El Niño, I'd give talks about them, but with hush-hush inferences to an anthropogenic link. One reason for my unfashionable (and non P.C.) climate silence regarding El Niño and the warming world were conflicts in the IPCC assessments regarding trends with hurricanes during El Niño cycles, which were reversed in later IPCC reports (something I'll discuss further when talking about Hurricane Harvey in the "Flood" chapter). More questions crept to the surface as Gore's rabble rousing climate campaign yielded to the often-fickle ratings that sell books and movies. Yet this wasn't a binary competition of who was right and who was wrong; climate change was real, but sounding the alarm too loudly had unintended consequences.

Adding to climate apprehension and El Niño doubt were subsequent IPCC assessment reports (like the AR4 and AR5) that later downplayed the gloom-and-doom that had become so fashionable from Al Gore's earlier climate pontification. Shortly after touring *An Inconvenient Truth*, it was revealed that Al Gore admittedly over-represented the case for climate change, rationalizing his distortions and magnified emphases as the best way to get people to listen. Most did, but some felt betrayed.[6] Other issues soon made Al Gore a questionable climate ambassador, like the time he stated during a Conan O'Brien interview on *The Tonight Show* in 2009 that the temperature of the interior of the earth is several million degrees – it's not, it's between 9,000 and 10,000 degrees Fahrenheit, something any earth-science student (let alone expert) should be aware of.

All the while, my talks and forecasts steered clear of slant and dealt with facts. I was told by more than one literary agency to change my tone since, "Slant sells." While that is certainly a great way to sell a story, I've found it can also result in ridicule and the inability to sleep soundly at night. Still, planetary changes were underway, and it's a shame that much of what Gore worked for became the butt of many jokes later. When truth be told, Gore had more things right than wrong — for what was known at the time. Yet he overstated a lot of things, and future reports by the IPCC never got attention as Gore seemed to keep less dramatic prognostications private. Even for what Gore got right, his delivery was a detriment, putting climate into question and his reputation on the chopping block.[7]

Unfortunately, unlikely ambassadors are appointed in the courts of public opinion, where they can also be tried, prosecuted and often persecuted for the perception of blasphemy. Similar to the case of Gore's admitted exaggerations, there became a sense of public distrust over the "*theys*" who said 2015-16 would be a "Godzilla" El Niño. It's a lesson that is not often learned by those with celebrity: *Be careful what you say, as you can taint the truth, and hinder its much needed dissemination.*

But that was then.

As 2015 approached and signs of a record-breaking El Niño developed early that year, I continued to qualify my forecasts, talks, and videos with caution, noting that no one can tell what will happen next, and that there are natural elements at play that will inevitably cause this El Niño to be like no other — yet in ways that could be, at first, quite disappointing. After all, not everything that the IPCC (or Al Gore) assessed came to fruition; in fact, opposing observations were being publicized including the infamous "global warming hiatus" — a deceptive stagnation from 1998-2013 when the earth didn't warm at all. That pause ended before 2015, right before a record El Niño was starting to take shape. Still, I felt it was a safe bet to straddle the fence of scientific reason and not lean my reporting to either side. I'm glad today that I didn't jump on the hyperbole bandwagon, but it was hard not to, given what was starting to unfold across the Pacific in 2015.

At the end of the day, the much anticipated, flooding rains didn't come during the El Niño winter of 2015-16. Sure, California saw its share of rain that winter, but it was nothing compared to what was about to come next, during the *neutral* Niño winter of 2016-17 — after El Niño had settled down.[8] But that's not how things looked in 2015.

El Niño's seemingly slow start in 2015 looked like a false alarm in many regards. After all, the media is not only known to hype things up, but, like in the 1970s when calling for an ice age, they can get it completely wrong. The media love to report end-of-days scenarios and they'll seek out anything that supports it — supposed ice ages in the 1970s or a warming planet decades later, it doesn't seem to matter either way, as long as it sells airtime. By 2015, climate advocates came under attack, not just for changing their minds from the '70s onward, but recently calling for things that simply didn't happen.

The most world-renown voice of climate change, who won the Nobel Peace Prize in 2007 for his efforts "*...to build up and disseminate greater knowledge about man-made climate change, and to lay the foundations for the measures that are needed to counteract such change*"[9] was, by 2015, being perceived as a false prophet having $30,000-per-month utility bills at his 10,070 square foot, 20-room home in Tennessee — a much higher carbon footprint than the people he was pontificating to.[10] And as seventeen ho-hum years passed since the momentous 1997-98 Big Friday El Niño, with no other major Niños hitting the Pacific, a "Godzilla" event seemed like another round of media embellishment, distortion, click-bait, or just down right hype. Yet the El Niño of 2015-16 was actually impetus for disasters yet to come. It was, in fact, just the beginning, sadly overshadowed by premature exhortation by the media. Patience would have been prudent — but that doesn't sell news.

The story of El Niño is a complex tale of not just science, but tragedy, excitement, and intrigue. It's a story of humankind's progress, yet precipitous peril. It's a story of lessons learned, and those we have yet to understand. It's a story about us, and how we need to respect the power of the Pacific and its influence across our planet. The story of El Niño though is more than just a story of surf; it's also a story about flood, fire, and deadly mud. Yet this story doesn't start with a science class; instead, it starts at the beach.

Backstory

Ocean is more ancient than the mountains,
and freighted with the memories and dreams of Time.

~H.P. Lovecraft

Although El Niño served as impetus to the recently devastating chain of events from 2015-18, it's something much, much older. El Niño is nothing new; in fact, there is strong evidence El Niño was very active during the Holocene epoch more than 10,000 years ago.[11] To put that in perspective, that's shortly after the end of the last ice age when humans started to realize the benefits of keeping edible plants close to home; or as we know it today: farming. Climate swings before and around that time — when wooly mammoths and sabre tooth tigers were entering extinction — didn't deter the human knack for survival. Instead of starvation in times of naught, our ancient, inventive ancestors figured out ways to keep their crops and survive through the harshest of times. Not all species were lucky enough to live through the last major ice age and the climate swings that brought them on; but we did. We adapted. El Niño and other climate cycles couldn't stop the progress of humankind, and many other forms of life on earth.

El Niños likely go back long before the last ice age as well, while earth's climate swings fluctuated along with it. However, it's only now, in what's being referred to as the *Anthropocene* (our recent epoch of significant human impact on the planet), that El Niños have become rather strong. But historically speaking, human ingenuity, coupled with the desire to out-do (or exploit) Mother Nature's swings between droughts, floods, famines, and times of plenty, has helped our species learn to live with El Niño. How we continue to adapt to infrequent climate cycles in a world changing from a larger, overarching climate change, is something that shouldn't necessarily cause panic, but prudent awareness. Our ancestors were smart enough to survive, but they may have paid closer attention to the world around them. Our ways of life today are different.

Shaman for millennia would foretell impending weather from the natural world around them. El Niño, past and present, is no different. If you were watchful of the ocean, its skies above, and the life above and below its surface, there was little need for orbital satellites, weather models or pressure maps to know if something was amiss. Some of the most notable ocean navigators, the Polynesians, passed on their knowledge of successful seafaring by songs, rhythmically telling their progeny of stars, birds, the speed of waves, and even the color of the sea and sky to know not only where to go, but what to avoid, and how to prepare. Viking raiders did basically the same thing, as did ancient Greeks and the warring factions told in Homer's Iliad, showing that people thousands of years ago could look at the sea, stars, and birds, and know how to get around. Peruvian fishermen had similar techniques as well, which eventually led to El Niño getting its official name.

As the story of El Niño goes, Peruvian fishermen, whose successful catches relied on their awareness of the sea, would at times feel the disappointing pains of meager catches of their prized anchovies, the Peruvian anchoveta. Today, Peruvian anchoveta harvests can yield an incredible eight or more tons of these heavily exploited fish.[12] To put that in perspective, with a single anchovy weighing just a couple of ounces, yearly harvests scoop up somewhere between 125,000 to 250,000 anchoveta each winter off Peru (more are harvested elsewhere around the globe). But sometimes, every few years or so, something strange would happen: The fish would leave, only to return a season or two later.

It didn't take long for Peruvian fishermen to see a pattern during the times of naught, but correlations are funny things, attributed only to what our current knowledge can piece together. Knowing the down times for anchoveta hauls happened only around Christmas, Peruvian sailors named this mystical phenomenon El Niño, translated from Spanish to English as "the boy", but when capitalized means "The Child Jesus"; thus labeling low fish-harvest seasons as omens of biblical proportions. No weather maps were needed; no surface temperature analyses were used; and altimeter readings for sea heights were non-essential. The old Peruvian fishermen knew all too well that when the anchoveta were gone, El Niño had chased them away. The fox had entered the hen house so the roost packed up and left. But what was *really* going on was much bigger than the Peruvian fishermen had imagined.

When Waters Warm

Anecdotal fish tales aside, there is a fairly simple science behind El Niño. It's a fascinating example of the tight coupling between our world's natural elements and how minor perturbations inevitably create transformations elsewhere on our planet. El Niño can drive the world's weather patterns and determine just how dry, wet, or even windy a region may be throughout an entire season. This is accomplished by seemingly minor fluctuations between the ocean and our atmosphere, which in turn rearranges earth's upper atmospheric wind patterns, known as jetstreams, and the storms they guide around the globe. El Niño's fluctuations from year to year and sometimes season to season, is like rearranging rocks in a stream of water, diverting the flow of things, which in turn affects other things downstream.

El Niño is, by its simplest description, merely a displacement of warm water that, instead of residing in the western equatorial Pacific, will sometimes move to the *eastern* equatorial Pacific. That's basically it. The cause for this is a bit more involved, but it's mostly just a matter of trade winds that continually blow from east to west across the Pacific, but every now and again those winds weaken. Those same trade winds were responsible for helping ships throughout history fill their sails, pushing them quickly along the equator from east to west. In fact, the Manila Galleons — Spanish trading ships that regularly sailed from Mexico to the Philippines from the 1500s to the early 1800s — used these east-to-west trade winds to make quick(er) work of crossing the Pacific Ocean, discovered in the early 1500s by none other than Ferdinand Magellan. After being blown at breakneck speed (on 16th century standards) to the Western Pacific, the return trip for a ship took advantage of winds located at a more northerly latitude, which, by the way, also helps to guide wintertime storms from Japan to Alaska, often sending swell to California's surf breaks. But the trade winds that blew ships in the 1500s from Mexico to the Philippines, being near the equator, has a major influence on our world's weather, and El Niño. What influences the east-to-west winds however, is a tiny trigger of Mother Nature's fickle Goldilocks-complex that just so happens to be linked to changes in our climate — water temperatures — which can't be too cold, or too hot, but just right to influence *normal* weather conditions.

Normally, the trade winds help to blow warm water across the equator toward the Western Pacific. But when those winds lighten up, a large pool of warmish water, which normally piles up on the western side of the Pacific from those trade winds, sloshes back toward the east, eventually setting up shop off Peru; hence, the tale of the Peruvian Christ Child and the mystery of the missing anchoveta. This mass of warm water is impressive, and deep, so when it slides back toward Peru, big changes occur. When that warm mass of water finally approaches Peru, El Niño is born (climatologically speaking).

The now-displaced, warmer waters off Peru start a chain reaction that shifts weather systems around the world. Warm waters cause low-pressure systems to form nearby, since air directly above warm water rises. During non-El Niño years, as air rises over the pool of warm water that's usually in the Western Pacific, cooler air flows in from the east to replace the void, thus creating the trade winds that blow from east to west. This provides a perplexing chicken-and-egg paradox: easterly trade winds blow warm water to the west, yet the trade winds are created by the warm waters in the west. If it not for one, the other would not exist.

Under normal conditions, the pool of warm water will sit along the western portion of the equatorial Pacific, and easterly trade winds will continue to blow. But when those trade winds weaken, even just temporarily, then the whole system goes into shock. The pool of warm water spills back toward the east, setting off a cascading, negative feedback loop where the trade winds, which are generated by heat rising off that pool of warm water, will continue to diminish since there is less warm water in the west to create the uprising of air for the trade winds to develop. This allows that pool of warm water to continue its abnormal easterly journey across the equatorial Pacific. This tightly coupled, interlinked, highly dependent, symbiotic system of warm water created by winds created by low pressure above the warm water (and so on, and so on) is a perpetual weather machine that sometimes breaks down. When it does, the pieces of this machine move along with it — wherever the warm water goes, weather shifts with it.

Since El Niño disrupts the position of low pressure systems across the entire equatorial Pacific, it also diverts storm systems around the globe. All weather systems around the globe touch each other as low and high pressure systems ride shoulder to shoulder along the planet's circulating winds and climate patterns. In normal years, storm systems in the North Pacific that would eventually travel to — and affect — the United States, usually travel along fairly high northerly latitudes, eventually bringing heavy rain to the Pacific Northwest, with lower amounts farther south. When El Niño though is underway, a large area of low pressure sets up a semi-permanent home in the Gulf of Alaska. This Gulf low, in turn, bends the jetstream to the south, changing the course for incoming Pacific storms. Instead of traveling to the Pacific Northwest, storms during an El Niño are more aligned on Southern California. As these storms traverse lower than normal latitudes across the North Pacific, they also tend to gain strength, kicking up stronger seas while tapping into copious moisture. The result is bigger surf directed at the west coast, and wetter weather for Southern California.

Other areas of our world are affected by El Niño as well. Heavy rains occur not just in Southern California, but also in Cuba, Northern Peru, Southern Brazil, Northern Argentina, Eastern Paraguay, Bolivia, and Western Europe. On the flip side, areas that would normally see rain are affected by droughts such as Indonesia, the Philippines, Southern India, Africa and Australia. These droughts in turn often lead to widespread brushfires, causing additional problems for these arid regions.

Yet El Niño has an interesting effect on hurricane formation, which at first can make this whole system seem counterintuitive. This though becomes a crucial piece of the story when we get to Act 2 — Flood — so a little background on this would be prudent at this point, before drawing back the curtain.

The Eye of the Storm

During an El Niño year, hurricanes are typically not as active in the Atlantic, but they are in the Pacific. During an El Niño year, the ocean waters in the Eastern Pacific are abnormally warm, giving hurricanes in this region more fuel than they're accustomed to. During the El Niño of 1997 for instance, there were a total of 19 named tropical storms in the Eastern Pacific compared to only 9 that year in the Atlantic. It's not often though that you'll hear about an active Eastern Pacific hurricane season, since most of these storms never come close to land. Nevertheless, 1997's El Niño provided enough high-octane hurricane fuel (warm water) to spin up, at the time, the strongest hurricane ever recorded in the Eastern Pacific: Hurricane Linda. Coincidentally, another Hurricane Linda formed during the El Niño summer of 2015, but it wasn't as strong as the 1997 storm. However, in 2015 the 1997 Hurricane Linda fell to the number two slot for strongest Eastern Pacific hurricane, being outdone by Hurricane Patricia later in the 2015 hurricane season, which peaked with astonishingly strong winds reaching a mindboggling 215 mph.

While El Niño keeps the Eastern Pacific hurricane season quite busy with warmer than normal waters, it has a unique effect on Atlantic hurricane formation. Since El Niño warps the jetstream in the northern hemisphere southward, it places this jet at a low enough latitude to blow not just into Southern California, but also across "hurricane alley" on the east coast. Having exceptionally strong winds, the jetstream blows hard enough to cause what's known as wind shear, which basically blows the tops off of rising hurricane columns, stopping these storms in their tracks. So while an El Niño often makes for an overlay active Pacific hurricane season, the Atlantic can become lackluster. But there's a catch.

The strength of an El Niño is measured by those warm equatorial waters in the Eastern Pacific. The higher the water temps are off Peru into the eastern equatorial waters of the Pacific, then the stronger the El Niño. But there can be a point of diminishing return. What determines an *effective* (not just measured) El Niño, is how the jetstream will be displaced. Warm waters off Peru, and the low pressure that moves along with it from the west to the east, also sets up a persistent low pressure system in the Gulf of Alaska that guides the jetstream into position. But there's no guarantee that the Gulf low will grow, move, or shrink, which in turn decides where the storm-guiding jetstream will be placed. Where the jetstream gets placed will determine what region will get wet, and see the biggest surf. So while warm waters in the Eastern Pacific can help fuel hurricanes for an El Niño summer, the winter is much trickier to call. Just because you have a strong El Niño signal from abnormally warm water temperatures measured in the equatorial Eastern Pacific, doesn't mean anyone knows *exactly* where the jetstream will be draped in the winter from the Gulf low forming in response to El Niño. We have a pretty good idea, but there is never a guarantee.

In the winter of 1997-98 the jetstream was smackdab over Southern California, and other Niño years were similar. But in 2015 we were dealing with a different beast. That El Niño measured higher than any previous Niño ever recorded. So would this make for a larger area of low pressure in the Gulf? Or would it shift the Gulf low to the east or west? Or perhaps the Gulf low would ride farther south? Any of these possibilities would determine a different placement of the jetstream, and subsequently different weather patterns as well. No one knew if the jetstream would be draped over SoCal, Baja, or maybe even just the Pacific Northwest — like during a normal, non-Niño year. How strong would the Gulf low be as well? Would it spin storms too quickly so that they didn't pick up as much moisture or have time to kick up massive waves from high winds? Would the Gulf low perhaps be broad but weak, thus putting the Pacific in a lull with barely any pressure difference between the Gulf low and high pressure nearby? And would the jetstream bend to a position over hurricane alley to put the kibosh on 2015's Atlantic Hurricane season? Entering uncharted territory with the strongest El Niño ever recorded made for a mystery. There were no simple answers — though some thought, incorrectly, that there were.

Going solely off statistics of equatorial water temperatures, as though all of the elements of El Niño would continue on some sort of linear progression — an assumption that a stronger El Niño would equate to stronger storms with no change in course from prior Niños — would be a mistake. But that's what forecasters were doing with the 2015 El Niño, despite numerous other signs that things were different in the Pacific that year. It's a mistake that, as I mentioned earlier, Bill Patzert from JPL seemed to make when he warned of a "Godzilla" El Niño with flood waters forcing you — as he put it — to get around your neighborhood in a kayak, not a car. Headline hungry media outlets played on this false assumption as well, increasing their ratings from mongering unfounded fear about things they did not fully understand. But in our intricate world of complex, meteorological independencies, where butterfly-effect kind of chaos systems rely on the sensitive dependence of prior conditions to determine future outcomes, it would be foolhardy to look into a cloudy crystal-ball that peered into a climate condition that no one had ever seen before; that is, unless one were brave enough to qualify their forecasts with a simple, "But we just don't know." Many meteorologists were cautious in that regard, but simple truth doesn't make for hot-selling headlines.

Things didn't turn out the way Niño alarmists had prophesied and the media, in turn, popularized. At first, things were somewhat lackluster during the El Niño winter of 2015-16 (compared to the hype at least), and less impactful than originally imagined. But later, things would get a whole lot worse. No one saw what was coming. El Niño outfoxed the forecasters.

Explosive Forces

The sea surface temperatures in the pool of equatorial warm water that determines an El Niño are staggeringly slight. During the intense El Niño of 1997-98, the maximum deviation in temperature measured around the Eastern Pacific was only 2.5°C (about 4.5°F) higher than normal when El Niño peaked around December of 1997.[13] In 2015, it was even higher, 2.6°C. As insignificant as this difference may seem, it's actually quite dramatic. By comparison, a one-megaton H-bomb would release enough energy to heat about one cubic kilometer of seawater by just 1°C. While that alone shows what a difference a degree can make, that's nothing compared to what it would take to heat the totality of the upper waters of the equatorial Pacific by 1°C the way El Niño does. This large expanse of water comprises a mind-boggling 6 *million* cubic kilometers of seawater, which means it would take millions of H-bombs to heat waters in the El Niño zones by just one degree. So when we compare the 0.1°C difference between the 1997 and 2015 El Niños, it's analogous to the energy of tens of — if not hundreds of — thousands of nuclear bombs.[14] The subsequent effect on weather is even more astounding.

The difference between the 1997 and 2015 El Niños was astonishingly large. Yet there was a complicating factor that no one had seen before: abnormally warm waters in the North Pacific, known technically as the *North Pacific Warm Anomaly*, but which many chose to call, the "Blob".[15] This was a result of El Niño's nemesis.

Four Years of Niña's Reign

The opposite swing from El Niño is known as La Niña, when trade winds become stronger than normal and blow too much warm water to the west, leaving the Eastern Pacific with cooler than normal waters. Being diametric to El Niño, where low pressure would dominate the Gulf of Alaska, high pressure monopolizes that region, which tends to block storms coming out of the Western Pacific, stopping them before making much in the way of surf to be sent to the west coast, while guiding rainstorms far to the north. The latter creates dry conditions for much of California, especially SoCal. So whereas El Niño sets up a storm track over Southern California, La Niña takes it away.

But there's another consequence from La Niña, especially when it lasts longer than normal. While the Gulf of Alaska is often dominated by low pressure, which is associated with stormy weather and cloudy skies, high pressure has the opposite effect, resulting in sunnier skies and less cloud cover. Having abundant high pressure occupying the Gulf of Alaska over an extended period of La Niña from 2010 until El Niño's return in 2015, provided an overabundance of sunshine and subsequent warmth to the waters in the Gulf of Alaska. This five-year period that hosted a dominant high pressure "ridge" (as these areas of high pressure are often called) in the Gulf was so bizarrely extended that it earned the nickname RRR, for "Ridiculously Resilient Ridge", coined by Daniel Swain, a Postdoctoral Fellow at UCLA.[16] This extremely resilient, ever-persistent area of high pressure refused to budge, and just stayed put — for years. The longer this high pressure ridge remained parked in the Gulf, the warmer those waters became. This led to abnormal warmth in Gulf of Alaska waters, and like all things in nature it too had a ripple effect.

By 2014 La Niña's extended reign of Gulf water warming led to what scientists were calling the "Blob" — a massive area of warm water that occupied a great deal of the North Pacific. It's believed there were other contributing factors that helped the Gulf heat up, including a lack of cooler water that would normally upwell, but with a shift in winds around that region less upwelling occurred to help regulate the Gulf's heat. There were also decades-long cycles at play, which I'll get to in the "Flood" chapter.[17] But no matter how you slice it, one thing is certain: warming waters in the Gulf of Alaska, before the kickoff to the 2015-16 El Niño, were now making more of the Pacific warmer than normal. Combine that with a stretch of equatorial water breaking El Niño's past temperature records, and you've got a mega-Niño on your hand. Signs of that became quite evident by the summer of 2015.

Surf

Waves are not measured in feet and inches;
they are measured in increments of fear.

~Buzzy Trent, big-wave surfing pioneer

When you drive to the beach and hear thunder, but there's not a cloud in the sky, you know something is wrong. When you next see wet roads yet there's been no rain, you realize danger may lie ahead. Then once you see walls of water many stories high marching toward a beach lined with yellow tape, police, and lifeguard patrols, most of us would get the heck out of the way and stay back. But not everyone does. In the middle of December 2015 though, the strongest El Niño ever recorded showed that you should.

December 2015 started out much like any other December along the California coast, but something was brewing more than 4,000 miles away that would change the face of El Niño in just eleven days. The winter surf season had started early that year with notable swells hitting Southern California around the middle of November 2015, and a variety of Western Pacific low pressure systems had created surf-worthy storms that brought significant swell to SoCal around the 4th and 7th of December. Those Pacific storms though were just the opening kickoff to a salvo of surf that would soon slam the California coast.

By early December 2015, anomalously high temperatures across the eastern equatorial Pacific reached 2.6°C above normal, almost 20% higher than the remarkable El Niño of 1997-98. A record-breaking El Niño was unfolding in 2015, yet this time around there was warm water across the *entire* Pacific with nearly all of the Northeast Pacific running 0.5-1.0°C warmer than normal; thus, the reason for the "Godzilla" epithet given to that year's El

Niño earlier. With plenty of warm water along the burgeoning winter storm track that was taking shape in the North Pacific, normally innocuous low pressure systems coming out of the Western Pacific were about to get a booster shot.

On December 5, 2015, four back-to-back areas of low pressure coming out of the Western Pacific hitched a ride on the jetstream, which by this time had lowered to a more southerly, El Niño-like latitude. Any storms driven along this storm track would make a beeline toward Central and Southern California, gaining strength along the way from this lower latitude, warmer water-fueled, storm-track trajectory. However, a small area of high pressure that commonly sits over SoCal was bending the jetstream to the north on the last leg of the jetstream's stretch across the North Pacific, thwarting would-be storms up and away from the SoCal region, within just a day or two before reaching the SoCal coast. However, much of this was about to change.

By December 8, 2015, all four areas of stormy low pressure coming out of the Western Pacific merged into one perfectly formed storm with enough wind to stir up 40-foot seas 2,000 nautical miles directly to the west of California's Point Conception — the dividing line between Central and Southern California. Swell, and lots of it, was now being directed at SoCal, calculated for a three-day, on-time arrival. There was no stopping it now; waves, big waves, were on the way. Swell energy in this distant Pacific storm had become so strong that waves being generated by this intense system would speed across the Pacific at a pace that would outrun the storm's weather front. Strong surf was headed to California; rain would then follow.[18]

By December 10, 2015, that small dome of high pressure over Southern California began to break down, widening a path for the swell-making storm to continue its trek toward the west coast nearly unabated. But the high was not absent; it had greatly weakened, but was still strong enough to bend the jetstream north at the last second, less than a day before the approaching storm would've had its chance to plow through SoCal. Yet although the approaching storm had settled down from its peak thousands of miles away a couple days earlier, by the 10th it was positioned at an uncomfortably close 1,000 nautical miles off the California coast, with strong seas stretching nearly the entire length of California. This not only brought surf, but by the 11th some rain had a chance to creep in as well, providing anemically low totals of about 0.1 inch of rain in many parts of Orange and San Diego Counties — lighter in Los Angeles. Almost none fell farther north in Ventura and Santa Barbara counties, which was puzzling.

Given that this El Niño storm was more wave than rain left many Niño watchers scratching their heads, since they were warned weeks before that biblical floods were on the way. Why this storm didn't live up to precipitable hope is something I'll revisit in the "Flood" chapter, with much greater detail. Nevertheless, while the meager rain that came to SoCal on December 11, 2015 was uneventful by El Niño standards, the surf that preceded it was monumental, to say the least.

On December 11, 2015, just shortly after midnight, remarkably powerful, 20-second period waves began hitting the buoys off California's Point Conception. Those periods don't just signify the time between wave crests; they are also a measure of strength and speed. Longer periods result in faster moving swell with eventually bigger breaking waves. While most swells that come to California in the winter have periods between 14 and 16 seconds, swell with 18 seconds is exceptionally strong. A swell with 20-second periods — like the one headed to SoCal — is inordinately intense.

By 3:00 AM seas had grown to 18 feet, and by 7:00 AM seas off Point Conception had grown to a jaw-dropping 25 feet. Waves the size of two-story houses were approaching the shore. By comparison, those same buoys on Big Friday 1998 topped out at 16 feet. The seas hitting Point Conception in December 2015 were 55% bigger. There was a slight difference though in that Big Friday's wave energy was a direct, westerly hit on Southern California from 275° whereas the December 2015 swell came in from a slightly more northwesterly angle of 295°. Once swells hit 300°, SoCal gets less wave energy, but swell on the 11th stayed just below that mark. This slight change of swell angle would save SoCal from a direct hit, but not by much.

During the early morning hours on December 11, 2015 the Ventura Pier was closed as it had once again sustained damage. Reinforcement efforts following the Big Friday swell of 1998 helped to maintain the pier's structure, but it was still no match for the 20-foot waves that were not just being swept *over* the pier, but traversing its entire length to shore. The normally laid-back Mondos — a popular spot in Ventura County with usually weakish waves, ideal for those learning how to surf — required a water rescue by the Ventura County Fire Department due to 15-foot waves breaking there.[19] And some Ventura residents were evacuated as waves crashed into nearby homes.[20]

Coastal flooding from the high surf was accentuated by King Tides on December 11th with high tides running nearly seven feet above sea level in response to a New Moon that crested that day. Harbor Boulevard, which parallels the coastline in Ventura was closed due to flooding as ocean water was thrust hundreds of yards farther inland than normal. High surf advisories were in effect for much of the California coast — not just SoCal, but Northern and Central California as well. Ocean Beach pier in San Diego was closed due to damage to the pier's railings, and incoming seas were strong enough to flip over concrete benches.[21] The parking lot at San Diego's La Jolla Shores was completely inundated with sea water.[22] Yet all of this flooding had nothing to do with rain; instead, it was waves.

The El Niño wintertime big-wave season commenced in California on December 11, 2015, but there was a sign one month earlier that surf would more than likely be rather robust in the months to come. In mid-November, the jetstream in the North Pacific was still riding at a fairly high latitude, more typical for a non-Niño year. That jetstream would lower in latitude in time to assist the December 11th event, just like it does during most El Niño winters. But being autumn, the jet remained at high latitudes when a large storm embedded itself in that storm track in mid-November. That storm stayed far to the north while riding high in the pre-winter jetstream, and ended up hugging the coast of Alaska. This storm though was able to strengthen into a frenzy of 40-foot seas as it approached the coast of British Columbia, fueled by the effects of El Niño — but not guided by it. For California, this storm's northerly course was bad news for surf as the incoming angle of wave energy was well past the 300° mark that, if lower, would've allowed heftier waves to make it to the coast — like the December 11th swell. But even with all odds of surfable hope working against the mid-November wave-making storm, it did send seas exceeding 20 feet by the time they hit buoys off Point Conception on November 16, 2015. Heavy surf in SoCal soon followed.

But it was the December 11th swell that told everyone in California that El Niño had arrived. It wasn't just a forecast of probabilities based on ocean water temperatures around the equatorial Pacific — it was real. El Niño was here, and things were about to get real.

The Biggest Wave

Other high-surf events followed the December 11th swell with one slamming Hawaii with such astonishing waves that all other swells in recent times can hardly compare. On January 13, 2016, with the jetstream being pushed into low latitudes by El Niño, a strong storm barreled out of the Western Pacific, riding the Niño-fueled storm track, which strengthened this storm rather quickly. Once north of Hawaii, blowing powerful winds in the 80 mph range, this system pumped up seas running an incredible 59 feet.[23] Even though this storm stayed hundreds of miles north of Hawaii while moving east, Hawaii got a decent, glancing blow from this mega surf-storm, sending massive waves to the north shores of the islands, while keeping the foul weather far from the region. Conditions would be epic.

On January 15th, outer water buoys off the Hawaiian islands showed seas in deep water running 30-35 feet.[24] As this intense wave energy approached Maui's northern shores, it ran into a shallower underwater ridge that signifies the subaquatic landmark of the world renowned, big-wave surfing spot known as Jaws. Located 30 feet below the ocean's surface, Jaws's underwater ridge interacts with incoming swells before the deeper water around it does, creating a process known as refraction that basically squeezes all of this swell energy together into a narrow band, resulting in behemoth wave formation. But when it comes to Jaws, "wave" is not the word that comes to mind; it's more like liquid cliffs that form cavernous tubes so large that you could drive train cars through them, without getting wet. Having such a deep water obstacle, 30 feet below the surface in the path of approaching wave energy, Jaws only reacts when something big — very big, actually — comes its way. When a 35-foot swell like the one on January 15, 2016 approaches Jaws, humongous walls of ocean water rise into gargantuan-sized, thunderously loud, shadow-casting, crashing combers. On January 15, when Jaws was woken up by prodigious ocean power, a legend of El Niño was made.

Jaws is most notable for tow-in surfers that rely on jet skis to bring them into a Jaws bomb, thus getting the speed needed to hydroplane the wave's face to ride it. But that wasn't the case when Aaron Gold made history January 15, 2016, riding the biggest wave that a surfer has ever paddled into and rode — no tow-in assist, just old fashioned arm energy to paddle, pop, then drop and ride.[25] Although bigger waves have been ridden by tow-in surfers, Gold's paddled-into wave measured 80 feet on its face, making Gold look like an ant riding a toothpick.[26] Although getting "rag dolled" at the end — a euphemism for getting violently thrashed once a wave breaks and mercilessly flops you around — Gold made history, and lived to tell the tale.

Shortly after that swell-making storm passed the Hawaiian longitudes, it quickly weakened and moved north, away from SoCal. Swell though hit the California coast two days later on January 17, 2016 with 14-foot seas tripping the Point Conception buoys with periods running 25 seconds — something almost unheard of. If you may recall, the December 11, 2015 swell had 20-second periods, which is a good 25% stronger than your normal wintertime swells in California. A swell with 25-second periods is so strong and fast moving that it's almost unworldly. This swell though didn't work out so well for California as rain got the best of it the next day. But earlier when it hit Jaws, history was made.

Niño y Nazaré

Meanwhile, in the Atlantic, another weather cycle had been brewing that just so happened to align with El Niño for big-wave timing, producing incredible waves at the home to some of the wildest weather in Europe. It's not too often that both the Pacific *and* Atlantic have simultaneous climate conditions ideal for big-wave formation, but 2016 was not your ordinary year.

Sitting off the coast of Portugal is a freak of ocean nature known as Nazaré, located on the edge of a town having the same name. Acting as a natural barrier against Vikings and pirates, Nazaré's cliffs and treacherous coastline provided a nearly impenetrable barrier against raiders for centuries — but not for today's big-wave surfers who wish to ride its castle-tall waves that resemble pyramids when they peak. While Jaws produces immense, blue, pristine waves that could win beauty contests, Nazaré is more like a monstrous ogre — massively sized, evil looking, and unfathomably powerful. Although not always churning up the prettiest looking waves, Nazaré has some of the world's biggest.

While Jaws provided Aaron Gold the world record for the biggest "paddled into" wave in 2016, Nazaré is where big-wave legend Garrett McNamara may have broken the Guinness World Record after he was towed into and surfed a 100-foot wave in 2013.[27] Although unconfirmed by Guinness currently, the video of McNamara's Nazaré ride shows an incredibly monstrous wave that seems unimaginable to surf — and live to tell about it.[28] But no matter how you slice the world-record pie, surf at Nazaré is, to put it simply, just freakin' huge. And like Jaws, it too went off in 2016.

Unlike Jaws's submarine ridge off the north shore of Maui, Nazaré's big-wave machine is a product of a deep underwater canyon that acts like a wave trap, luring in deep water energy that would have otherwise lost its oomph when entering shallower waters near the coastline — a process known as shoaling. Having deep water to flow through as swells fast approach the shore, almost none of their wave energy is lost before suddenly hitting shallow water close to shore, forcing profuse quantities of water upward into immense, breaking waves. Nazaré though has more tricks up its sleeve.

On either side of the Nazaré deep water canyon, as swell energy is being focused into the canyon at high speeds, slower moving wave energy joins forces with the canyon-drawn wave energy, creating a much bigger amalgam of wave energy in a process known as constructive interference. As these combined swells finally hit the end of the canyon and focus this multiplied energy on the Nazaré headland, prodigious mounds of water rise to form some of the biggest waves in the world, with peaks that completely overtake the horizon. Like fluid mountains rushing toward the coast, Nazaré's waves build to nearly 10 stories tall.[29] But conditions have to be right for the Atlantic to gin-up enough swell energy to get the ball rolling. In February 2016, it did.

The world's oceans are riddled with cycles. El Niño is just one of them, which is sometimes complemented with a bigger pattern in the Pacific that also happened in 2015-16, which I'll cover in the "Flood" chapter. But the Atlantic has its cycles too, which don't always align with — or are caused by — El Niño, but have profound effects nonetheless. One such cycle is known as the North Atlantic Oscillation, or just NAO for short. This cycle, while not necessarily having a link to El Niño, has a lot to do with the waves at Nazaré. Besides having a record-breaking El Niño, coupled with the freakishly warm North Pacific "Blob", the winter of 2015-16 also saw a positive NAO pattern.[30] It were as though the surfing stars aligned, providing a world of big-wave surfing in not just the Pacific, but Atlantic as well.

The NAO winds-up when there's a great difference in pressure between a permanently placed area of low pressure over Iceland that interacts with a permanent area of high pressure to its south known as the Bermuda-Azores High — something I'll revisit later when talking about wild hurricanes that affected the Atlantic the following year. On February 17, 2016, those areas of pressure were strong, which in turn created a tight pressure gradient between them, which in turn made for exceptionally strong winds with a weather pattern conducive to surf-worthy storm formation. As these pressure masses strengthened and rotated against each other, their opposing interactions spun-up a hefty storm that kicked up seas in the North Atlantic measuring over 30 feet, which then marched toward the shores of Nazaré.

Although losing steam when the storm neared the coast of Portugal, with seas running about 12 feet by then, the swell periods were still long enough to exacerbate the effects from Nazaré Canyon's, wave compression cannon, which not just doubles the size of strong, long-period swells, but quadruples them. On February 19, 2016, 50-foot waves slammed the coast of Nazaré as surfers including Mick Corbett, Jarryd Foster, David Langer, Sérgio Cosme, and others braved the monumentally sized mountains of water head on.

Headlines lit up across the surf community that El Niño was bringing big waves to Nazaré, which wasn't *technically* true. While El Niño is caused by immense areas of equatorial water sloshing from the Western Pacific to the east, which in turn shifts weather-making areas of high and low pressure systems, the NAO, which helps to kick-start the Nazaré wave machine, is more a matter of two permanently placed areas of atmospheric pressure that just expand, contract, and/or shift slightly from side to side from time to time, with no causal link to shifting water temperatures from El Niño. In fact, if you were to overlay timelines to compare positive Niños and NAOs, there is hardly any correlation — not enough to say there actually is. Also, in February 2016, the North Atlantic had exceptionally cold waters, measuring 2°C below normal in some places.[31] When the NAO went positive again a year later, those same Atlantic waters were much warmer, once again showing no clear correlation between El Niño and the NAO; however, early 2016 was a time when multiple forces of the world's weather and climate joined forces — when it rains, it pours...so to speak.

A serendipitous coincidence brought the big-wave surfing community a smorgasbord of goliath-sized waves around the planet in early 2016, which in itself raises the eyebrows of those monitoring weather, waves, and climate. But back in the Pacific, El Niño was doing its thing.

Titans and Bellwethers

Shortly before the explosive surf-fest at Nazaré, another swell, fueled by El Niño, hit the California coast late January into early February 2016, slamming the buoys off Point Conception with seas nearing 20 feet. Although a stormy swell, it, like the major swells before it, got an El Niño assist. But it wasn't until later in February 2016 that the annual, big-wave surfing competition known as the Titans of Mavericks would be held. Drawing in big-wave competitors from around the world to congregate outside Pillar Point Harbor, just north of Half Moon Bay, California, challengers would soon surf the infamous, notorious, and sometimes deadly spot known as Mavericks.

With the green light given in early February 2016 that the contest would soon be underway, the Titans of Mavericks competition was on — waves were coming. Swell was wintertime big for this Northern California contest on February 12, 2016, but not epic in size. Seas hitting the Point Conception and Northern California buoys that day topped out at just 13 feet, not the 18 feet from the December 11th swell, nor 14 feet from the mid-January swell. Those other days were questionable for conditions with light rain and southerly winds. February 12, 2016 was practically pristine for a big-wave surfing event with no rain and light offshore winds for the better part of the day. While surfers love El Niño, it doesn't feel the same way. El Niño oftentimes brings weather along with its waves — something we'll see more of in the next chapter.

Mavericks can push relentless waves that punish those who attempt to tame and ride them. Sitting in the cold North Pacific waters off Northern California, this big-wave break focuses energy from wintertime swells onto an underwater ramp, flanked on both sides by faster-moving energy in deeper water. Being almost the inverse construction of Nazaré's wave machine, Mavericks produces waves with similar, astounding results. But it is hardly for the faint of heart, as unlike the warm waters of Jaws off Maui, Mavericks greets its takers-on with brutally cold, often brown-ish looking waves, colored by seafloor sediment stirred up by the incredibly strong, and often incessant breaking waves that disperse along a rugged and rocky coastline. Surfers who dare to challenge a wave at Mavericks are reminded of its ominous lair as gravel sometimes spits at their backs from the churning, crashing waves closing in behind them. Like a wild, bucking bronco, Mavericks has a way of telling its riders they shouldn't be there. But there are those who successfully tame the beast.

Nic Lamb of Santa Cruz, California, won the Titans contest in 2016 but others weren't so lucky like Ken Collins, who had to withdraw after nearly drowning from a wipeout while riding a 30-foot wave-face.[32] Although Nic Lamb got the top spot, everyone who surfed Mavericks on February 12, 2016 was a winner; if you survived, that was your prize. Collins and Lamb were lucky. Other surfers had ridden their last wave at Mavericks including Mark Foo who dropped into an 18-foot wave too late, wiped out, and drowned in 1994. Sion Milosky drowned at Mavs in 2011 after being held underwater way too long while incessant waves refused to let him surface after a wipeout. At "Ghost Trees" surf spot to the south of Mavs, Peter Davi drowned in 2007 after trying to paddle into a wave that, like others earlier that day that reached 70 feet, was powerful enough to kill.[33]

Intense wave energy raised a virtual surf-titan from the deep starting in December 2015, lasting well into 2016 — in time for the contest at Mavericks. Surf though is just the first sign of things to come from an El Niño, acting as a kind of canary in a coalmine. Unless you spend a lot of time on the shores of the west coast of the U.S., this auspice of El Niño goes unnoticed. Walls of water approaching the shore are just facades, hiding the approaching storms that sent them.

It's hard to wrap one's head around stormy weather ahead when sunny skies complement big surf days. Smiling faces of hootin' surfers getting their fill of stoke riding strong, Pacific waves can lull us into thinking that all is well, even though it is not. Similar to tsunamis where deceptive outflow of tidal waters is a warning to get to higher ground, high-surf events along the coast of California tell us conditions are changing far, far away — changes that will soon arrive, and then stick around for a few years with myriad aftermath.

Throughout this story of *Surf, Flood, Fire & Mud*, I'll revisit how waves served as signals during seasonal changes. Surf, in many regards, is a bellwether leading the way for the remaining three acts that round-out our story. When waves come ashore, they tell stories of storms from far-off places that stirred up seas many thousands of miles away, and many days prior, from disturbances that may have started many months before. As a matter of fact, although there was colossal surf pounding the coast of California on December 11, 2015 from a storm that formed about a week earlier many thousands of miles away, El Niño had arrived long before then.

El Niño broadcast its arrival many months before California saw surf; not just by showing it had warmed water temps in the eastern equatorial Pacific, and not just by doomsday, "Godzilla" headlines earlier that year. Instead, it was through observations that you could see for yourself on a quiet stroll on the sand. If you visited a beach in California during the summer of 2015, you could clearly see signs that something was amiss.

SHARK!

While breaking waves begin the story of El Niño's arrival, that's just the tip of the proverbial iceberg. Under the waves there lies a deeper story — figuratively, and literally speaking. Surf is merely the manifestation of ocean disturbances, but those same, overarching effects can be powerful enough to shift the entire food chain. When that happens, humans are no longer the top apex predator near the shore.

Bruce Brown, in his famous, classic documentary *Surfing Hollow Days* said it best that "...sharks have a bad habit of eating people." Referring to Australia, man-eating sharks are so common down-under that it sparks debate on the need for a series of 50-some shark nets around New South Wales, coupled with a battalion of drum lines strewn off Queensland. Australia faces a difficult balance between high shark- and people-populations, resulting in environmental collisions from unintentional bycatch deaths of dolphins and sea turtles in the shark-deterring nets.[34] California though (especially SoCal) doesn't face such problems, and is usually a fairly shark-free region — until El Niño.

Overall, California isn't known for the abundant sea life it once saw. Many fisheries collapsed along the California coast in the 1950s, mostly from overfishing. Numerous species continue to struggle to make a comeback today. In a sense, much of the California coast is a dead-zone, at least in comparison to what it was before canneries became commonplace in the 1940s, leading to the eventual decimation of natural fisheries. Like a tombstone marking the death of an era, there now sits a single stretch of hotels and tourist attractions along Monterey's Cannery Row, once the most popular of cities during the boon days of California's fish industry, whose over-fishing demise was popularized by Steinbeck's book of the same name.

When the fish left California waters, so did the things that ate them. While there are still plenty of sharks around the waters off California, the survivors have adapted to more of a seal and sea lion diet, keeping most of the larger sharks in deep waters, especially around the Channel Islands in Southern California where the dog-like, sea mammal prey congregate with little to no threat from land creatures — only those in the deep ocean.

Surfing in California for many decades now, I can anecdotally attest that there are (usually) very few sharks around SoCal's beaches. I've only seen a few in all my years of surfing, although I'm sure many more were nearby. Still, they tend to avoid any kind of human interaction (humans have a bad habit of killing sharks, a man-bites-dog kind of irony when you think about it). While some regions of SoCal — like portions of the Santa Monica Bay — are known as pupping grounds for young great whites, these nurseries are home to smaller sharks that would never dream of attacking something close to their size. Besides, SoCal is known to be a fairly shark-safe surfing zone with only 36 confirmed, unprovoked shark attacks recorded since 1926.[35] There have been even fewer fatalities of surfers and swimmers in that tally, including a great white that killed David Martin while swimming off Solana Beach in 2008; a mako that killed Michelle von Emster while swimming off Point Loma in 1994; a great white which killed Tamara McAllister while kayaking in 1989; and yet another great white that killed Robert Pamperin while free-diving in La Jolla cove in 1959. These shark attacks are rare compared to Florida (with over 733 attacks since 1926), Hawaii (143 attacks), and especially Australia (591 attacks). Southern California is a place known for non-aggressive sharks — a relatively safe place to swim, surf, and kayak. But the El Niño of 2015-16 changed the playing field as ocean temperatures shifted, and the food-chain followed suit.

On September 5, 2015 as ocean waters were starting to respond to El Nino's shifting temperatures, Dylan Marks was fishing from his kayak about a mile off shore near Deer Creek Road in the Malibu region, located just south of the Mugu Rock — a semi-natural structure made famous as a backdrop for many car commercials showing a coastal drive. Marks had been dangling his feet off of his kayak in the warmer than normal waters off the coast of SoCal — something he surely regrets today. After reeling in a surprise catch, a hammerhead — usually found much farther south off Baja — the shark found Marks's foot to be either appealing, or just something in the way of escape.[36] Either way, shark bit man. Leaving the gory details to your imagination, Marks survived after an air-lift and numerous stitches. But this was only one in a series of hammerhead encounters that took many Southern Californians by surprise.

Nearly two weeks later, another kayaker had a close encounter with a hammerhead, shooting a video of the incident that went viral. Bill Morales, like Marks earlier, was fishing from his kayak, this time a bit farther south off Dana Point. Placing his catch alongside of his small watercraft was probably not a good idea, given that word had already spread about an influx of hungry hammerheads into SoCal waters. In doing so, Morales's catch attracted a hammerhead that would not leave him alone to fish in peace. Instead, his catch was ripped to shreds as the hammerhead refused to vacate, obviously ravaged by hunger, and desperate enough to chance an encounter with a human-piloted kayak to fight for meager scraps of surf-catch. This was, according to Morales, his fourth such encounter in the past few weeks; in fact, in another incident that Morales filmed, a hammerhead circled his kayak for quite some time before giving up and swimming away; that video too went viral.[37]

The hammerhead incidents with Marks and Morales were surprising, but there was already news of hammerheads around the area, including reports by divers who had spotted large schools of them, something usually seen in Mexico, not California. In fact, on August 14, 2015, a 58-year-old man was bitten while diving off Cortes Bank, 100 miles off the coast of San Diego — an indicator that this shark species was migrating north. A 6-foot hammerhead was also caught off the Santa Monica pier in August 2015, and there were reports of other hammerheads circling fishing boats searching for tuna. An influx of hammerheads into SoCal became common knowledge in late 2015; in fact, while I was giving a talk in Ventura that summer to local business men on the subject of El Niño, many of these gentlemen (who also participate in various water sports) told stories of hammerhead schools observed in the northern reaches of SoCal waters, in the Santa Barbara Channel.

Hammerheads had come to Southern California, something that was very hard to believe, and easily brushed off as old men telling fish tales. But these were not fictive stories of braggadocio; the unthinkable was happening, and many stayed out of the water for a while.[38]

These were things that you just didn't see around Southern California, but El Niño's warming waters, after scattering schools of fish from their frequent feeding grounds, left predatory fish to fend for themselves, hungry, and desperate to find food. But this was only the beginning.

Ocean water temperatures were taking longer than normal to recuperate from the El Niño of 2015-16. On May 10, 2017, video shot from an Orange County Sheriff's Department helicopter spotted 15 great white sharks swimming curiously near paddle-boarders near the coast of Capistrano Beach.[39] Water-goers were already on high alert after a shark attack at nearby San Onofre State Beach in April of that year. But as El Niño continued to wane, water temperatures eventually found equilibrium by 2018. The El Niño of 2015 though had other surprises in store.

Snakes on the Strand

Silver Strand Beach in Oxnard, California has been home to a popular surfing break for decades. Having a reputation decades ago as a locals-only surf spot, "The Strand" is highly coveted by nearby residents who enjoy its clean breaking waves. The Strand was also home to a hardcore surfing community (from back in the day) which spun off the 1980s Nardcore style of punk music with bands like Dr. Know, False Confession, Stalag 13, Scared Straight, and Ill Repute. If you surfed the Ventura Coast in the '80s, it was common knowledge that unless you knew someone at The Strand, there was a high probability that your car could become the victim of smashed windows and/or slashed tires — a deterrent to try to keep "kooks" from stealing waves from the locals. I still remember many a time seeing the Oxnard Police mobile unit parked around the area on weekends — a paddy-wagon to provide an express lane for handling the high weekend crime volume. Things have changed over the past 30+ years, but the one thing Silver Strand Beach — being almost 200 miles north of Mexico — is not known for, is venomous sea snakes. Even the far southern reaches of SoCal are void of them, but like hammerheads and anchoveta, snakes were also on the move during the El Niño of 2015, and some headed far to the north — to The Strand.

On October 16, 2015, Robert Forbes, while walking the beach at Silver Strand got quite a surprise when he spotted an exotic (and incredibly venomous) yellow-bellied sea snake.[40] An earlier sighting by Anna Iker a couple of days earlier may have been the same snake, although no one will know for certain since the snake from her sighting washed back out to sea. Robert Forbes's sighted snake though did not get away, and in fact, was taken to the Los Angeles Museum of Natural History. Sadly, the trip to the museum was one for post-mortem, as this snake had passed away while waiting for state wildlife officials to arrive on scene after being called by Forbes earlier that day.

About two months later, on December 21, 2015, another yellow-bellied sea snake came ashore near Huntington Beach, California, discovered during a beach clean-up event by the Surfrider Foundation.[41] Interestingly enough, a few years later in January 2018, another yellow-bellied sea snake came ashore in Newport Beach in Orange County, California, measuring 25 inches in length, discovered near the 18th Street lifeguard tower. Although El Niño-zoned waters had cooled by then, they continued to be abnormally warm in many regions of the Pacific in early 2018, including a path for sea snakes to take from Mexico to Southern California.

Emaciated, starving, and in very poor health, the rare, interloping, venomous sea snakes in 2015 were just a mere fraction of literally millions of foreign sea creatures that would find their final resting place on California's beaches. Displaced by hunger, confusion, and navigation to parts unknown, life in the ocean came under attack by a force of warm water with a Spanish name.

Crabs

As news of the strengthening El Niño spread in the early summer months of 2015, San Diego bore witness that El Niño was more than a headline. Sea life in SoCal, overall, was off-kilter, which you can expect to some degree during an El Niño. Past El Niños, for instance, brought in record numbers of albacore caught off California in 1997-98.[42] Dusky sharks — a species that rarely travels farther north than Mexico — were also spotted in SoCal during the 1997-98 El Niño. Guadalupe cardinal fish became more commonplace in SoCal around 1997-98. And while Pacific barracuda are no stranger to California, Mexican barracuda are, and were caught in late 1997 around Oceanside and La Jolla, California.[43] Unexpected catches like these are common from an El Niño, but 2015 brought enigmatic marine events of far greater magnitude.

In June 2015, San Diego residents substantiated the seriousness of that year's impending El Niño as millions of red tuna crabs washed up on their beaches. So thick were the masses of these tiny red crabs — many millions in all — that residents had to clear paths through them or else walk on a crunchy carpet more than a foot thick of dead and dying crustaceans.[44] The abundance of these pelagic red crabs is normal, but where they ended up was not. Traveling in massive schools, the red tuna crabs meander within the currents and moderate water temperatures that normally flow far off San Diego, which provide the plankton that the tiny crabs feast on. But as El Niño shifted the water temperatures in the Pacific, it affected the plankton due to a lack of upwelling — where ocean currents lift life from the ocean floor up to the surface, and then back down again, in a circle-of-life kind of way. With upwelling disturbed around the California coast, the food chain was perturbed into partial collapse, leading the tuna crabs into a death spiral of starvation — and likely confused navigation — eventually beaching themselves, *en masse*, as a last-ditched, yet fatal effort for survival.

Tuna crabs have an El Niño commonality to the Peruvian anchoveta in that small perturbations to water temperatures shift their life in very big ways, in turn creating a ripple effect throughout the upper levels of the food chain that depend on this food source. Many fish find tuna crabs to be a regular staple, including billfish, yellowtail, grey whales, blue whales, sea otters, and also sharks. But as millions of tuna crabs carried out their final journey to their sandy graves on the beaches of San Diego, their swansong sojourn left numerous predators wanting, forcing billfish, yellowtail, whales, sharks and more to select things other than these tiny crabs for their seasonal sustenance. The crabs went absent and their predators migrated to unknown regions to find their next meal.

Slugs

Of all the sea life die-offs and displacements witnessed during the El Niño of 2015, one of the more disgusting events was the washing up of giant, purple sea slugs onto Northern California beaches. Leaving a snail-trail of dark purple ooze on the beach next to where they lay dying (or dead), these exotic, mucousy, marine invertebrates portrayed a rather grotesque sight for passersby. Beachcombers stumbling across the dead slugs thought they were witnessing remnants from a violently gruesome murder scene and called 911.[45] The number of victims was not great, but the view was shocking.

Also known as sea hares, these sea slugs are not foreign invaders to California waters; however, the massive die-off was unusual, the second such mass die-off in the 15 years leading up to the 2015 El Niño. As news crews were already attentive to marine life disruptions due to El Niño in 2015, the sea slugs made headlines with pictures that, if not for the headlines, could send shivers down your spine – or at least make you a bit queasy.

Accompanying these larger blobs was another species not so indigenous to Northern California. As early as February 2015 there was a population explosion of tiny southern sea slugs making their way into the normally cool (but now warmer) waters off San Francisco. While common in San Diego and Los Angeles, these tiny, colorful slugs are quite uncommon north of San Luis Obispo. Making their way into waters 200-400 miles north of their normal home, the southern sea slugs were adaptable El Niño survivors — they followed the warmth. These smaller sea slugs are a great example of evolution in the making, where those with predispositions for adaptation will live long enough to pass on their genetic structure; while those that can't, won't. At the same time, although their story ends better than other creatures affected by the 2015-16 El Niño, the tiny sea slugs serve as yet another reminder that something in the waters was just not right.

The Dead Zones

It can be difficult to relate to problems taking place thousands of miles away. Watching our beloved pets age and pass-on is far more relatable than thinking about schools of anchoveta being displaced, sharks struggling to find food, crabs dying on beaches, or sea slugs finding their way to confusing locales, dead on arrival. Even the poster child of climate change, the polar bear, wasn't always enough to sway the minds of humankind until it was anthropomorphized by t-shirts, commercials, and stuffed animals, bringing a cute kind of humanity to these creatures so that our empathy for them would turn into action to save them. Proximity is everything when it comes to empathy.

Yet when it comes to relatability, unlike Las Vegas, what happens in the ocean doesn't just stay there; instead, it follows us everywhere, evident by El Niño's cycle causing the *en masse*, sea life displacements. In a way though, these are sucker punches in the battle for earth's sustainability — we're up against the ropes, and El Niño just keeps pounding way, exacerbating the strain on already, highly stressed ecosystems. So while we look through weather-colored glasses when viewing El Niño and disastrous events it later unfolds on land, it's also good to peer at what lies beneath those warming waters.

Earth's oceans are in sad shape compared to just a few decades ago. According to the World Wildlife Fund, marine species populations are half of what they were in 1970. In less than a century, humankind has decimated creatures that occupy the better part of this planet.[46] Given that we're 65 million years past the last great extinction that killed off the dinosaurs, which allowed mammals like us to evolve and flourish, humankind has, in just the last 0.0000001% of our recent evolutionary trend, been able to wipe out 50% of all life across the entire world's oceans. A good deal of this decimation is due to commercial fishing, where the world's taste for seafood, sushi, shark fins, mussels, clams, crustaceans, and even Peruvian anchoveta, is draining a limited supply, with not enough time for replenishment, especially during natural, periodic climate cycles like El Niño.

It's like we never learned a valuable lesson: Although human farming came into play after the last ice age over 10,000 years ago, to this day we act like pre-historic hunter-gatherers in the world's oceans — a practice we learned ten millennia ago that would never lead to survivability. These were lessons learned all too well (and all too late) centuries ago by now defunct (or completely missing) peoples such as the Easter Islanders and Greenland Norse; where over time they realized that without careful resource management and monitoring you are merely taking steps to collapse your entire civilization.[47] But that doesn't seem to stop us, and El Niño can make it worse, delaying recovery.

As El Niño was underway in 2015, further stressing sea life, Japan continued to ignore recommendations from the International Whaling Commission (IWC) and was on the hunt, yet again, to slaughter over 300 minke whales in the Southern Ocean – all in the name of what, they say, is science, but is anything but.[48] Many trinket shops in China, lacking now-rare elephant tusks to carve and sell their versions of scrimshaw, have resorted to harvesting giant clams, resulting in devastating impacts on the local ecosystem as these reef-dwelling giants — that take up to 100 years to reach incredible girth and weigh up to 400 pounds — are now, too, being removed from the world's oceans.[49] Humankind, more than El Niño, is affecting our planet, making it a perilous place to maintain an accustomed way of life. Not planning on El Niño, which can exacerbate the issue, makes it worse.

Besides over-fishing, there are other anthropogenic factors responsible for species decline, including pollution from run-off and, more frightening, a change in the acidity of our oceans due to increasing concentrations of greenhouse gases that circulate through the cycle of ocean-atmosphere exchange. Everywhere we turn, the *advancement* of humankind is taking its toll on the wildlife inhabiting Earth's oceans. So when we talk about the effects of El Niño on fisheries, and all the bizarre displacements evident as El Niño shifts ocean temperatures, currents, and weather, it's important to understand that these amplifications are due to earlier influences of humankind. We're seeing these events unfold through a magnifying glass, observing proverbial final straws straining the backs of ocean wildlife. And the strain has been great, especially since El Niño started its comeback in 2014, broke records in 2015, and showed its aftermath in 2016, 2017, and 2018.

During the El Niño that kicked off in 2015, Chile saw unprecedented marine-life die-offs. In December 2015, more than 10,000 squid washed ashore in Chile, for reasons unknown. More than 330 whales also washed up on beaches in Chile during the year of this record El Niño.[50] Also in 2015, in one incident, 8,000 dead murres (a kind of half penguin half sea-bird species) were found on the banks of Whittier Beach, Alaska, starved to death due to prey displacement linked to El Niño.[51] At the same time, more than eight times more fur seals than normal were found stranded on the California coast, most emaciated, and nearly half of them dead.[52] These die-offs are attributed to a shift in the food chain from El Niño — the same attribution given to the millions of dead tuna crabs and missing anchoveta, setting in play a domino effect that reverberated through the food chain, sparked by El Niño. And now we as humans, the topmost apex predator and dominant species on earth, are not just the last link on that chain; we're also the first.

What Comes Up

The surf zone in California is known for murky water, but not the crystal clear, Caribbean-like conditions witnessed many times during the El Niño summer of 2015. The extended visibility in SoCal waters in 2015 was welcomed by surfers, swimmers, divers and sport-fish enthusiasts. Yet at the same time, nature's unclouded aqueous beauty off the beaches of California was a sign of just how brutal El Niño can be. While the clear waters brought smiles to the faces to those entering the surf-zone off the California coast in the summer of 2015, it was a deceptive siren; a beauty with hidden dangers; a menacing manifestation of El Niño upending life on our planet — literally.

Ocean currents provide vital nutrients to feed the web of life, and the California coast is a prime location for a phenomenon known as upwelling: a top to bottom circulating replenishment to keep the virtual lifeblood of the ocean in good health. Standing at the shores of the ocean and looking out over the vast sea, it might be hard to comprehend the myriad motions working the waters to and fro. Breaking waves provide proof that what goes out tends to come back in — or in the antithetical action through outflow and rip currents. Yet our world's oceans have a variety of currents circulating salt, heat, and nutrients through these bodies of water, as though the oceans themselves were a form of life, not just an environment for it.

These ocean currents provide impetus to a butterfly-effect that, with a mere flap of a metaphorical wing, can drastically change the outcomes of elements linked upward and outward through its long and trailing chains. But unlike the famous light-wing-coined concept of chaos theory — which was originally going to be called "the sea gull effect" but the more attractive *butterfly* supplanted the label — the currents in the ocean are not chaotic, yet can be disturbed, especially during an El Niño, leading to the bizarre events covered in this chapter.

If you've ever watched *Finding Nemo*, you've seen an animated version of a popular ocean current known as the Gulf

Stream. While this fun, animated movie showed sea turtles surfing a fast moving current, in reality it's a bit slower than what was depicted, but the concept is still the same. The East Australian Current, depicted in *Finding Nemo*, is just one of many legs of the thermohaline circulation, which acts as a main artery for the ocean's central circulatory system. In the northern hemisphere, a large current also circulates on the boundaries of the oceans in a clockwise rotation, known as a gyre, responsible in large part from the rotation of the earth and blowing winds (gyres exist in the southern hemisphere too, but they rotate the other way). Both currents work in a synchronized dance: In the Atlantic, the clockwise rotating currents draw warm water from the tropics northward along the east coast of the U.S. But in the Pacific, California is on a colder leg of that ocean's current, where cold water is driven from Alaska south along the west coast of the U.S.

While gyres may be astounding, the thermohaline portion of the cycle (the current exploited by Nemo and his newly friended sea turtles) is an even more amazing display of how earth can balance itself into equilibrium: As warm water is drawn north to the Arctic it cools and eventually freezes. Since only the water freezes, salt is naturally extracted, which then drops toward the ocean floor (a process known as *brine exclusion*). The falling salt pushes water down with it, starting a current that eventually forces the water back toward the equator, where the process begins again. These gyres and thermohaline streams are the reason why Virginia Beach has warmer water than San Francisco, despite being on the same degree of latitude — Virginia receives warm water traveling north from the tropics, but California receives cold water flowing south from the Gulf of Alaska.

But the wetsuit-required waters off much of the coast of California are cooled even further from upwelling — a phenomenon that California was perfectly built for, yet is vulnerable to climate swings like El Niño. Northwesterly winds, common to California, blow against the coastal mountain ranges that line the coast of the Golden State. This guides the winds into a parallel position along the coastline. And then, as the planet rotates, ocean surface water is effectively pulled away from the beaches by these winds from an effect known as Ekman Transport. The warmer water pulled away from the coast is then replaced by colder waters upwelling from the deep. This upwelling is the reason why many swimmers and surfers are often shocked at how cold the surf zone can be from one day to the next, all depending on how windy it was the day before (strong winds one day upwell colder water by the next day). But most importantly, this upwelling stirs up sediment from the ocean floor, causing the murkiness of California's coastal waters, while also bringing nutrients to the surface — like plowing a field in preparation for planting.

Phytoplankton at the surface, energized from the sun and the upwelled nutrients, are swarmed upon by zooplankton, which in turn is fish-food for species like Peru's anchoveta, and the sardines that are now mostly gone from the waters off California's Cannery Row. Bigger fish and mammals, like seals, then feast on the smaller fish, and then apex predators (sharks and humans) feed on them. Upwelling is genesis to the great circle of life, starting with wind that blew for an extended period, causing the water to upwell and begin life anew. If though the wind pattern changes, then this mechanism collapses, as it did during the El Niño of 2015-16.

In 2015, water temperatures were shifting wildly across the Pacific, causing a change in atmospheric pressure above the water and the subsequent winds they produce. The disruption of wind patterns caused the normal upwellings to change in frequency, thus causing the waters off the coast of California to become, in a way, stagnant. But unlike a pond or lake, where stagnation is evident from stench and murky surface filth, California's upwelling-less stagnation resulted in a partial collapse of the food chain. El Niño brought clear, warm water to the California coast, but it was not necessarily a thing of beauty; it was a sign of death in the coastal waters — a symptom of severe, ecosystem collapse.

The Cyclone Signal

Besides the intense surf in December, the summer of 2015 was exceptionally active with hurricanes and typhoons across the Pacific, with a number of Northeastern Pacific hurricanes bringing surf to Southern California beaches. It's something that raised a signal that El Niño was afoot, yet an inverse kind of event led to a different surf scenario in 2017 — something I'll get into more in the "Fire" chapter when talking about "The Surf Signal" leading up the firestorms that year. But back in 2015, with exceptionally warm, North Pacific waters, hurricanes and typhoons had plenty of fuel to work with.

All told, in the Northeastern Pacific, there were 26 named storms, 16 of which became hurricanes, outdone only by the 1992 season, which had 27 named storms at the tail end of a moderate El Niño season.[53] In the Western Pacific there were 18 typhoons, 9 of which became "super typhoons" (cyclones that equal the strength of Category 4 or 5 hurricanes). At one point in August, Hawaii was surrounded by three strong, Category 4 storms: Kilo, Ignacio and Jimena — a very rare event.[54] Most of the time hurricanes will spin off the warm waters near Costa Rica and drift westward, sometimes with a nudge north to direct swell at Southern California. But most of the time these storms just drift toward Hawaii and fizzle before hitting the islands. And, they usually don't form simultaneously, as was the case with Kilo and friends.

While hurricane activity is less active in the Atlantic during El Niño — due to stronger winds over hurricane alley that literally blow the tops off of forming hurricanes — the Pacific does just the opposite and fires up storms from abnormally warm water (and a virtual absence of wind) during hurricane season. In 2015, the Pacific was plenty enough to warm to fuel the plethora of hurricanes and typhoons that year, but there was a strangeness to these storms from the bizarre paths they took. While there was warm water everywhere, and in many cases, not where one would expect during an El Niño (something I'll dig into deeper in the next chapter), storm tracks for the 2015 hurricanes and typhoons were quite peculiar. This was no ordinary El Niño, and is wasn't just the *number* of hurricanes that rang the alarm — their treks did as well.

Olaf, for instance, after becoming a Category 4 storm in October 2015 headed on the usual westerly route, but then turned around and started to make an easterly beeline for Southern California (and northern Baja). Hurricanes Andres and Blanca that started out the season took a warm-water path north toward Baja, stopping just short of entering SoCal waters. Dolores in July, another Category 4 storm, took a trek up the Baja coast, eventually fizzling out right before crossing *into* SoCal waters. Other storms were able to drift north into warmer than normal waters as well like Kevin, Linda, and to some degree Sandra, but the oddest of storms was Ignacio, which moved first toward Hawaii, and then north into the Gulf of Alaska before turning back to the east toward Canada — an extremely rare trek for a storm that depends on warm water to stay alive. Hurricane Oho did something similar, forming near Hawaii and then moving north into Alaska — a place where hurricanes just don't belong.

The 2015 Pacific hurricane season was one for the record books — not so much for the number of storms or intensities, but how these systems behaved and the courses they took. Weather patterns across the Pacific were quite different than ever before, but meanwhile, in the Atlantic, there was a contrast.

As you may recall when talking about hurricanes in the "Backstory" chapter, El Niño summers place the jetstream over the Atlantic's hurricane alley, creating high winds in the upper atmosphere that shear off the tops of rising hurricane columns. So during an El Niño year, while the Pacific is bathed in warm water with few obstacles to thwart forming storms, the east coast of the U.S. has lower hurricane activity. Such was the case in 2015 when only four Atlantic hurricanes formed, none of which reached category 5 status, and only one (Joaquin) reached category 4 status. In fact, only one other storm was semi-strong, Danny, which reached category 3 status.

Pacific weather maps though were lit up like Christmas trees during the entire 2015 hurricane season. Since the Pacific was firing off storm after powerful, cyclonic storm, it had to mean something. Thoughts of climate change (not just climate cycles) came to the forefront.[55] Was the abundance of Pacific cyclones a signal of a changing climate, El Niño, or both? There's an interesting distinction when trying to answer that, since they are, in many regards, mutually exclusive in regards to hurricane causality. Correlations are indeed funny things, since not everything is always directly connected.

Although climate change was (of course) underway, none of the experts were — or are even now — linking 2015's hyperactive Pacific hurricane season to global warming. Armchair pundits, alarmists and activists may say differently, but the ones who study the science to subsequently set the record straight, say otherwise. In fact, a recent summary by NOAA[56] urges caution to connect hurricane activity to climate change, stating it's "premature" to link human causes to tropical cyclone activity; however, NOAA also says that climate change could, at some point, have an effect — but none are detectable right now. Additionally, the latest IPCC report (AR5) states there is "low confidence" that any changes in tropical cyclone activity are linked to any kind of human influence, and that natural causes play a major role (something I'll touch on again when talking about Hurricane Harvey later in the "Flood" chapter).[57]

Intuitively, your gut instinct may tell you that this does not make sense and it goes against the grain from what cyclones were signaling in 2015. But if you hear *all* of what the experts say, it does. Cherry-picking only certain climate signals draws attention away from natural elements and cycles that occur no matter what we do, so it's important to draw distinctions between the hand of humankind and the inner workings of the natural world. Doing so we can better mitigate problems within our control while realizing it can exacerbate those that aren't. There's rarely a case where we can divide onus to these issues into either column A or B. But when it comes to hurricanes, the line of distinction is clear: there is no link between human-induced climate change and hurricanes. However, El Niño falls on a precarious division.

While climate change is a reality, not all of what happened over the 2015-18 span of surf, flood, fire and mud, can be *directly* linked to it. In fact, a number of natural elements started to come together prior to 2015, in some ways repeating history from events that took place in the 19th century from some very unique, long-term, natural climate cycles that change the state of our oceans (and subsequent weather) over many generations. Occurring so infrequently, it's hard to realize what these cycles of change are until they're knocking at our door, ready to come ashore. But what's more important, is that these cyclic changes don't just affect surf — waves are just the opening bell. Climate cycles can take tens, hundreds, and even thousands of years to circle back around. And if, by chance, these cycles, each on varying timescales, happen to align their lucky stars and come together all at once, then hold onto your hat — the ride will get wild, and often deadly.

The cyclone and surf signals in 2015 were reflective of these climate cycles, which started long, long ago. These cycles have been so pronounced at times that throughout history they've affected the outcome of battles, wars, and even the progress of the American West. Nature is powerful in more ways than we realize. We're merely guests on a ball of dirt and water that zooms around a star at 67,000 miles per hour. The planet we live on, which has been around for billions of years and seen multitudes of lifeforms rule for millions of years before going extinct, cares not for our wellbeing. We may romanticize about our connection to Mother Earth, but the wrath of her nature tells us she cares less for us than we do her. Instead, earth merely reacts to its place in the cosmos, from how it spins, wobbles, tilts, and otherwise interacts with the star it orbits, which in turn affects its oceans, land, and ice, and the subsequent cycles of weather that ensue. These cycles, in turn, change us.

Next time you're at the beach and you see (or ride) a wave, revel in the fact that the wall of water you observe, no matter its size, is an omen sent from a storm far from where you stand, created by elements of the cosmos that came together at a given point in time, telling of eventual change that may, some day, change the world in ways we may not yet conceive. The climate cycles that drive our weather and waves are a part of who we are; they've occurred many times before; they continue to occur today; and they will be with us tomorrow.

But they will not always work in our favor.

Flood

Until the latter fire shall heat the deep,
Then once by man and angels to be seen,
In roaring he shall rise,
And on the surface die.

~The Kraken, Alfred Lord Tennyson

California sits on the razor's edge between hell and high water. A dichotomy exists between scorching deserts and the world's largest body of water, separated by a thin line of inhabited soil and shifting sand. There's a tragic irony that excessive heat and fire sometimes occur just feet away from an overabundance of cold, Pacific water. But that large ocean, spanning from California to Japan and Alaska to Antarctica, with its temperatures quite disparate from those over land, can whip up some of the wettest storms in the world. And it doesn't have to wait for El Niño; in fact, it's what sometimes happens *after* an El Niño that can have the greatest consequences.

The El Niño winter of 2015-16, despite the hype, was a bit of a bust for rain. Oxnard, CA, which on average gets 14 inches of precipitation during its rainy season from October into the spring of the following year, received only 8 inches of rain during the 2015-16 rainy season. Yet a year later, after El Niño had calmed down, Oxnard saw a deluge with 20 inches of rain — 40% higher than an average rainy season. Northern California had similar results with Sacramento measuring only 16 inches of rain during the El Niño winter of 2015-16 (about 3 inches lower than normal), but then Sacramento saw nearly 30 inches of rain the next winter — a more than 50% increase from average, *after* El Niño had already passed. Los Angeles followed suit with only 9 inches of rain in the winter of 2015-16 but then nearly 17 inches of rain in

2016-17. And farther south, San Diego had below average rainfall measuring just 9 inches in the winter of 2015-16 but then 13 inches the following, neutral Niño winter.

Although major surf pounded the California coast in the winter of 2015-16, the scant precipitation that fell that season made laughing stocks out of the "experts" who were calling for a "Godzilla" Niño. Bill Patzert — a research scientist at JPL who coined that name earlier in 2015 for the impending El Niño — took center stage in the news that year, adding other comments to stoke the fire of flood-fears including saying in December 2015 that,

> *There's no longer a possibility that El Niño wimps out at this point. It's too big to fail.*[58]

He also called for a doubling of California's normal rainfall.[59] But as rain was late in coming, Patzert stubbornly held on to his El Niño guns. Responding to a CBS reporter in December 2015 when being questioned if this Niño was really going to live up to the hype, Patzert defiantly said,

> *Everybody is impatient here in Southern California, 'where's the El Niño'. Sit tight. In January and February, you'll essentially be riding your kayak in front of your home.*

With confidence comes calm; but insecurity breeds unfounded defiance, and dogma. Science can be doubted; that's the whole idea: present your theory and if it doesn't hold up after being questioned or results say otherwise, then adjust, move on, and try again. The voice of scientific reason should handle comments to the media with a bit more tact.

While Patzert was obstinately preaching the second-coming of Niño and the end-of-days that would follow, reaching for rationalization to support his earlier, doomsday, headline-making, prognostications, other scientists had reservations. Mike Halpert, deputy director of NOAA's Climate Prediction Center wouldn't go out on the doom and gloom limb, but instead called for the *potential* for a wetter-than-average winter, while cautioning that "when you're dealing with climate predictions, you can never get a guarantee."[60] Chris Dolce, from The Weather Channel was cautious as well, saying,

> *There is no guarantee that we will see overwhelming amounts of rain in Southern California in the months ahead. It's impossible to pinpoint just how wet and frequent the storm systems that may affect the region this winter and early spring will be.*[61]

Halpert and Dolce, while playing it safe, were wise to do so. But ho-hum discretion and prudent projections do not make for click-bait, shock-and-awe, eye-popping, attention-grabbing, sponsor-funded headlines. With media competing for every second possible in 24-hour, 7-day-per-week news cycles — where Kardashian gossip spars with presidential tweets to get viewers' undivided attention — hyperbole won the day. More of the media focused on Patzert's sky-is-falling forecasts than the level-headed Dolce's or Halpert's. With more news covering Patzert's kayak-through-the-streets scenario than the possibility that *we just don't know what the heck will happen*, information (false as it was) about El Niño spread quickly, warning of an impending, overabundance of Noah's Ark-like rain storms. But that's not what happened.

After the relatively dry winter of 2015-16, with many giving dismissive waves of their hands and turning their backs on "experts" calling for catastrophe, concern returned to a common and very serious issue that often faces the American West: drought. After all, for a number of years prior to this hopeful period of El Niño rain, California suffered one of the worst droughts in its history. A few years earlier, the winter of 2012-13 saw only 5 inches of rain in Oxnard; 2013-14 not quite 6 inches; and just a tick over 7 inches in 2014-15 when El Niño was starting to gear up. These amounts were less than half the normal 14 inches that Oxnard gets during an average rainy season. Given the extended lack of rain, reservoirs were being depleted. Most of California's lakes in 2013 had reached historic lows, some down to just 30% of their capacity. By October 2015, all of California, Oregon, Washington state, Nevada, and parts of Idaho, Utah and Arizona were in drought. The majority of California was in the worst stage of drought known as a D4; the rest of the state was between D2 and D4 status, signaling severe drought throughout the entire state.[62]

Water rationing became mandatory in most of California by 2014. And on April 1, 2015, as that year's El Niño was gearing up, Governor Jerry Grown, by executive order, instated mandatory water-use reductions for the first time in California's history. After four straight years of severe drought, the state's resources were over stressed. At a news conference that day, Governor Brown stated,

> *People should realize we are in a new era. The idea of your nice little green lawn getting watered every day, those days are past.*[63]

I never quite understood what Governor Brown meant by that tone. Obviously he wanted citizens of California to reduce their water use — I get that. But the *way* he said it seemed demeaning, coming from the state's head official, as though having a patch of grass on the expensive yet small plots for homes in his state was a bad thing. To say "nice little green lawn" seemed like either a childish taunt, or a snobbish, ivory-tower means to a water-shaming end. Brown makes other bizarre remarks later, that I'll touch on in the "Fire" chapter. But *how* he said the little-green-lawn remark isn't as important as the context I'm sure he wanted to convey: We're in drought; it's here; get used to it.

This sparked concern of rising food prices and a possible population exodus from the state. Finger pointing between residents and farms that use much more water, helped stoke the anger of the dry-weather weary. With California farms using 80% of all the water in California, Governor Brown's little-green-lawn comment seemed slanted, leaving many homeowners frustrated.[64] Nevertheless, many homeowners gave up with the idea of two-day-a-week watering and just turned their sprinklers off completely; some opted for desert landscapes instead, others just let their yard turn into patches of dirt. Leading up to — and especially following — the "dry" Niño of 2015-16, California residents resigned themselves to a new normal: water was, and would continue to be, a rare commodity. New revelations concerning California rain took hold: Rain wasn't coming; the forecasters got it wrong; and Governor Brown was pissed-off about little green lawns. All but the last of those ideas would change — in time.

With the media hungry for ratings-worthy headlines, quotes from Bill Patzerts over-ambitious forecasts brought hope; however, it was followed by disappointment. When El Niño finally came to fruition in 2015-16, rainfall was unimpressive compared to the strength-signal of that year's record-breaking El Niño, which measured 2.6°C above normal in the ocean waters around the eastern equatorial Pacific's El Niño zones. This measurement of anomalous sea surface temperatures is used as a way to measure the strength of a Niño: the higher the temperature signal, the stronger the Niño. By comparison, a strong El Niño in 1982-83 had a lower signal of 2.2°C and saw about 30 inches of rain in Oxnard, CA — more than twice the normal amount. A similarly strong El Niño during World War II, which started around 1940, had a peak signal of 2.0°C, which brought almost 25 inches of rain to San Diego. That WW II El Niño, by the way, which was more than 20% lower than the 2015-16 Niño, had much broader implications.[65]

In fact, the 1940 El Niño, which lasted for a few years, was powerful enough to partially change the course of World War II. By 1941, when that era's El Niño was gaining strength, German troops were advancing into the Soviet Union on the notoriously bloody Eastern Front — where eventually 27 million Soviets and 4 million German troops would be killed. In the summer of 1941, things originally went along swimmingly for the Germans as they made progress toward the Soviets in hopes of further conquest, but they were stopped in their tracks (literally) from exceptionally harsh winter weather. In the early 1940s, El Niño had flip-flopped the more usual, expected weather patterns that war tacticians normally plan for. While Alaska saw exceptionally high temperatures, there were times during the harsh winters over El Niño's multi-year stay, that temperatures on the Eastern Front dropped to -68°F.[66] Earlier, Joseph Goebbels, Hitler's Minister of Propaganda, broadcast a desperate appeal to the people of Germany for warm clothing to send to the troops, as the exceptionally harsh winter had overstressed their supplies, even after looting from Russians and Poles. Martin Bormann, the head of the Nazi Party Chancellery, frustratingly (and stupidly) remarked,

> *One can't put any trust in the meteorological forecasts.*

Bormann, thinking his racist, German-pure heritage put him on a pedestal above reproach, believed he was more of a meteorological expert than those providing weather forecasts for the war. Bormann is quoted as also saying,[67]

> *Weather prediction is not a science that can be learnt mechanically. What we need are men gifted with a sixth sense, who live in nature and with nature— whether or not they know anything about isotherms and isobars. As a rule, obviously, these men are not particularly suited to the wearing of uniforms. One of them will have a humped back, another will be bandy-legged, a third paralytic. Similarly, one doesn't expect them to live like bureaucrats.*

Hitler's war machine, filled with arrogance, hubris, hyper racism, intolerance and downright ignorance, lost the war. And El Niño, at one point helped the allied cause. That momentous point in history was affected by an El Niño that had a signal of about 2.0°C — something neither Bormann, Goebbels, or Hitler saw coming. Stronger El Niños would come generations later.

The 1997-98 El Niño, which brought the colossal waves on Big Wednesday and Big Friday, had a signal of 2.4°C, which resulted in over 36 inches of rain in Oxnard — two inches of that, by the way, fell in May during a bizarre, late season storm when El Niño was going into a temporary neutral state (a surprisingly rainy condition that I'll be elaborating on throughout this chapter).

But the strongest El Niño on record, in 2015-16, with a record-breaking signal of 2.6°C, saw *much* less rain — *below* normal, in fact. It was truly pathetic, easily beat by one of the weakest El Niños in 1994-95, which had a barely qualifying signal of 1.1°C, yet brought 20 inches of rain that season to Oxnard, CA (about 40% more than normal). Nevertheless, in the winter of 2015-16 the rain didn't come. It eventually would, but certain elements of nature in 2015 were conspiring for delayed reaction.

The PDO

El Niño has a bigger brother known as the Pacific Decadal Oscillation, or PDO for short. While El Niño cycles occur every few years or so, the PDO cycle shifts ocean temperatures on much longer time scales; hence, the decade part of its name. The PDO also affects a much larger area of ocean water than El Niño does; it doesn't just warm equatorial waters, it warms (or cools) entire sections of the Pacific. Similar to El Niño and La Niña, the PDO has warm cycles and cool cycles — they just span much larger ranges and timescales, and subsequently have greater impacts on the world's weather.

During a PDO warm phase, much of the eastern edge of the Pacific — from Alaska to Chile — experiences anomalously warm water. That expanse of warm water also includes Peru and the infamous, equatorial El Niño regions of the eastern Pacific. So when El Niño *and* the PDO are in warm phases, El Niño tends to be amplified from the addition of warm PDO waters into the warm El Niño zones in the eastern equatorial Pacific. When the PDO goes warm, El Niños have a better chance of occurring, and becoming stronger.

There's a correlation to this, which is especially evident from two major El Niños: the Niño of 1982-83, and the Big Friday-causing Niño of 1997-98. During both of these El Niño events the PDO was in an exceptionally warm cycle. The opposite holds true for La Niña years, which occurred recently with a cold PDO cycle that started late in 2005 and ended by 2014, during (and exacerbating) the drought that preceded the 2015-16 Niño. And then, right on cue, as the PDO entered a warm cycle late in 2014, like a synchronized dancer staying in lockstep with its climate partner, El Niño returned in-kind.

By 2015 the PDO was well into a warming cycle. Throw in the El Niño forming then and you are starting to get the makings of a powerful El Niño. But there was one more ingredient that would be added to the climate cauldron cooking up El Niño in 2015, which ended up throwing a temporary wrench into the weather works.

The Blob

Things were really heating up in 2015. As you may recall when discussing the backstory earlier, the North Pacific in 2015 was warming from not just El Niño and the PDO, but also from another, bizarre anomaly that many were calling the "Blob"— known technically as the *North Pacific Warm Anomaly*. While it's been debated as to whether the cause brought effect or perhaps the effect brought cause, studies show a correlation to a massive area of warming waters in the North Pacific in 2015 and the years of La Niña that preceded, eventually helping to shift the climate cycle in 2015.[68] Years of sunny skies over the Gulf of Alaska from the prolonged period of La Niña from 2010-14, coupled with a cool phase of the PDO at that time, helped to maintain La Niña's lengthy reign. This resulted in the normally cool waters in the Gulf of Alaska warming abnormally. This was thought to be adding more fuel to El Niño's fire, which makes sense at first: El Niño means warm water, so more warm water means more El Niño...right? Well, not necessarily.

Three elements in 2015 conspired to provide abnormal warmth in the Pacific: El Niño, a warm PDO, and the abnormally warm Gulf of Alaska waters from the Blob. One could then theorize that the El Niño forming that year would bring copious rain. Statistically that makes sense, if you derive that idea on a measure of water warmth (the warmer the Niño the rainier it would be in California). After all, abnormal warming led to myriad sea life displacement discussed earlier. And if one correlates water warmth to past Niños, one would draw a rainy conclusion for California. But correlations are funny things.

Weather on our planet is a result of relativity, not just statistical correlations. Since ocean waters are cooler and more consistent than air temperatures are over land, and since the equator is warmer than the poles, there are differences of temperature across the entire planet. Each area of temperature encompasses an area of pressure: high pressure has sinking air (from cool temps) and low pressure has rising air (from warmth). When the differences in temperature — and subsequent pressure — are great, weather starts to form. But the operative word here is *difference*, something lacking in 2015.

During the mega El Niño of 1997-98, while equatorial, Niño waters were exceptionally warm, the rest of the Pacific was not. In fact in 1997, while the equatorial El Niño waters ran 2.4°C above normal, most of the Pacific ran a good -1°C below normal. In other words, there was a wide range of disparate water temps on a scale spanning 3.5°C (from -1° to 2.4°). But in 2015-16, although equatorial El Niño waters were a record-breaking 2.6°C above normal, most of the Pacific was warm as well, at least 0.5°C above normal across most of the Pacific outside of the Niño zones. This meant that the range of disparate water temps in 2015 was spanning only 2.1°C (from 0.5° to 2.6°). In other words, the 2015-16 Niño had less difference in water temperatures across the Pacific, by as much as 1.4°C compared to the 1997-98 El Niño.

This is one reason why many forecasters approached the El Niño winter of 2015-16 with caution. No one knew for sure what part the Blob would play that season. The Blob was something new

— a black-swan kind of event, something never seen before. There were no statistics to compare it to. No El Niños had ever encountered the Blob before. And with an exceptionally warm El Niño being backed up by a warm PDO cycle, the Blob became a climate enigma. There were many questions: Would there be more moisture in the atmosphere from a larger area of warm water evaporating in the Pacific? Would the warm waters thwart the idea of a low pressure system setting up in the Gulf of Alaska that would be typical for an El Niño? And would there be enough difference in temperatures across the North Pacific to support strong storm formation from pressure differential? No one knew for sure.

A lot was learned during and after the El Niño of 2015-16. Bill Patzert retired from JPL a couple years later — not because his Godzilla expectations turned into paper tigers that went flat; instead, after 35 years working at JPL, at 76 years old, Patzert decided to move on to quieter, greener pastures.[69] Many who took his Godzilla hype to heart realized that sometimes people get it wrong; after all, with just two years to go before retirement, after more than three decades of arduous work in his field designing weather satellites and watching the world's weather, his heart might not have been fully in the game, especially when a possibly annoying journalist rubbed him the wrong way on a not-so-good day.

Scientific research though continued after the Blob and Niño 2015 passed quietly into the history books. Studies continued to try and understand what caused the Blob, what role it played in weather and El Niño, if it could ever return, and if it did, what would happen. To this day though, like so many things climate- and weather-wise, despite knowing what we do, there is so much more we don't. The scientific community can honestly say, we're just not completely sure. But history tells us we better find out.

History's Lesson

With fear of flooding on the minds of many Californians in 2015, there was concern that history could repeat itself; not so much from 1997-98, but also the 19th century. The largest flood in the recorded history of California (and Oregon and Nevada as well) occurred back in the winter of 1861-62, when new settlements in the American West were unaware of the Pacific's history for occasionally record rainfall over the region. Rain pelted California almost continuously from December 1861 through January 1862, causing extensive, catastrophic flooding. The heaviest rain fell during a 30-day period from December 23rd through January 22nd, with almost 26 inches of rain falling in some areas of Northern California. In fact, a two-day span starting on January 10, 1862 resulted in 11 inches of rain — in just two days alone.

In Los Angeles, where less rain tends to fall compared to northern portions of the state, it rained non-stop for 28 days in the winter of 1861-62. The mining town of Eldoradoville, that once sat near the San Gabriel Mountains, was completely washed away — and was never rebuilt. So much water flowed through the usually dry riverbeds around LA that the Santa Ana River merged with the San Gabriel river, forming a complete expanse of water covering a distance of about 18 miles from Long Beach to Huntington Beach.[70]

Unprepared for storm drainage and rain management, the "Great Flood" of 1861-62 resulted in extensive loss of livestock, with 100,000 sheep, 500,000 lambs, and 800,000 cattle destroyed by flood waters. Up to one-third of California's property was destroyed as approximately one in eight homes was carried away by flood waters, or was completely ruined. Oyster beds around San Francisco were decimated from an excessive influx of freshwater entering the bay from all the rain runoff. Almost every part of California was touched by the Great Flood of 1862.

Other states were affected by the flood as well including Oregon, Idaho, Nevada, Utah, Arizona, New Mexico, and even parts of Mexico. But devastation to the "Golden State" was the worst — and it wouldn't be the last. What caused the Great Flood is something that ended up affecting weather in California one year *after* an El Niño had passed — a pattern that would repeat itself more than once.

It may be hard to fathom it, but El Niño wasn't to blame for the Great Flood of 1861-62. Moderate El Niños occurred years earlier in 1857, 1858 and 1860, but by the winter of 1861-62, the El Niño cycle was neutral.[71] The recently wet winter of 2016-17 was similar: El Niño occurred the prior winter, but the wet winter of 2016-17 was during a time of neutral Niño. That's not to say that winters are wettest a year after an El Niño event; instead, it's a matter of having conditions that are more conducive to rain than drought, which happens quite often during a prolonged period of a neutral Niño state following an El Niño. After an El Niño, it can be common to see the opposite swing occur, La Niña, which tends to be a drought-maker. But when Niño goes neutral and doesn't swing wildly either way, conditions can align for a perfectly prime period of extended rain on the west coast of the U.S.

Something similar occurred in December 1964, known as the Christmas Flood, which was the worst flood in recorded history on nearly every major stream and river in coastal Northern California. While not affecting as broad of an area as the Great Flood of 1862, the Christmas Flood brought copious amounts of rain to the Pacific Northwest, with Crater Lake, Oregon receiving 38 inches of rain in December 1964 — 26 inches more than normal. The Christmas Flood ended up killing 19 people. It devastated 10 towns, caused heavy damage to 20 highways and bridges, and killed 4,000 head of livestock.[72] Like the 1862 flood, December 1964 was not in an El Niño, but instead was in a neutral Niño state — sandwiched by a weak El Niño the year before, and another the year after the flood.

Weather and climate cycles in 1862 and 1964 have eerie similarities to the recent events starting in 2015 lasting into 2018. A severe drought had struck the U.S. during the 1850s, estimated to be was worse than the Dust Bowl in some parts of the country.[73] The recent drought in the American West preceding the 2015 El

Niño was also followed by flooding rains in 2016-17, after an El Niño. But the most poignant parallel between the 1862 & 1964 floods and the record-breaking, rainy California winter of 2016-17 was a weather pattern that can tap into the most copious components of perceptible moisture in the Pacific. These are known as atmospheric rivers, and when Niño goes neutral these patterns can turn preeminent, and deadly. Such is the story of the Great Flood of 1862 and the Christmas Flood of 1964, when atmospheric river storms brought rain of biblical proportions to the American West and Pacific Northwest. History would repeat itself by 2017.

When Atmospheric Rivers Flow

A storm is like a canon in that it needs ammunition to be effective. In the case of low pressure systems that drive weather fronts, moisture is needed to fire-off a rain event. You can have the lowest of low pressure systems, but if there's no moisture nearby for it to tap into, then there won't be any rain. But, if there *is* moisture for a low to tap into, then that system needs to move slow enough so that it has time to stir in the moisture, and eventually spit it back out as rain. Two things are thus necessary for a successful rain event: moisture, and slow storm speed. When these come together, flooding often follows.

While the Pacific Ocean has an abundance of water, there isn't always a lot of evaporated moisture above it. But when warm tropical waters have cooler air above them, air then gets pulled upward, taking ocean moisture along with it. The cooler it is in the atmosphere above the warm waters, the faster this moist air rises. In hurricane regions, during the warmest times of the year for ocean waters, this disparity between warm water and a relatively colder upper atmosphere is great — another example of weather and temperature relativity. When that difference in temperature is excessive, hurricanes begin to take shape from rapid rising columns of warm, moist air. But when these columns are weak, you get something a bit different.

A steady flow of warm, moist, rising ocean air occurs almost regularly near Hawaii and southward into the tropics, which establishes a large plume of tropical moisture that rides along the Hawaiian latitudes. As this moisture meanders and establishes a moisture plume in the atmosphere, it becomes a sitting duck for any storm that would wrap its rotating tendrils into it. But to be effective, a storm needs that plume to be drenched with moisture. When El Niño isn't bending the jetstream's strong winds over lower than normal latitudes, tropical moisture plumes have a better chance to rise, grow, and amble about. That's when things can get dicey.

When one of these moisture plumes becomes big enough and a rotating area of low pressure bumps into it, an atmospheric river can form. It's like a fork twirling a plateful of pasta, where the fork is the low pressure system and the pasta is the moisture plume. During extreme seasons of Pacific atmospheric rivers, large areas of low pressure indicative to the Gulf of Alaska drift toward the west coast of the U.S., which often just sit and spin off the coast. As one of these stormy, rotating lows sits off the coast, its counterclockwise rotation eventually draws up tropical moisture plumes to its south, eventually forming a steady flow of moisture to feed this storm. The longer that low can sit and spin, the longer the atmospheric river continues to flow.

The Great Flood of 1861-62 is a prime example of this, when an atmospheric river pattern stayed put for over a month, feeding in a continuous stream of flooding rain to California. Same goes for the Christmas Flood of 1964; an atmospheric river pattern set up over the California-Oregon border, streaming in what seemed like a non-stop supply of rain to that region. The 1964 event was especially primed with atmospheric moisture as that year saw the most active Pacific typhoon season ever recorded with 40 storms, some of which lasted into December of that year.[74] Both of these great flooding events though didn't occur during an El Niño; they happened after.

During non-Niño years (especially neutral Niño years) atmospheric rivers can be enhanced. El Niño, as far as the west coast of the U.S. is concerned, is all about dropping the jetstream to a lower latitude to drive Pacific storms along an unabated path to Southern California. If we enter a La Niña, then that jetstream is bent far to the north, resulting in drier conditions in SoCal as storms stay clear of that area. But when El Niño slips into a neutral state, a few things happen that can enhance these atmospheric rivers.

During a neutral Niño (like we saw in the wet winter of 2016-17), the jetstream stays at a moderate latitude to allow storms to travel across the Pacific — not lined up on SoCal as it would be during El Niño, but not completely torn apart like in a La Niña. The jetstream also tends to become less persistent during neutral Niño years, allowing storms to move slower, and in some cases stay put. The transport of tropical moisture to feed the atmospheric rivers' moisture plumes is also different during neutral Niño years; in fact, studies have found that neutral Niño years have more tropical moisture feeding atmospheric rivers than during El Niño years.[75] This jives with rainfall totals in neutral Niño years that followed El Niños.

Take for instance the El Niño in the winter of 1991-92, which brought 14.5 inches of rain to Los Angeles. Once that Niño quieted down the next year, things went into a neutral Niño state over the winter of 1992-93, which ended up bringing 22 inches of rain to LA — a 50% increase from El Niño the year earlier. If though El Niño quickly shifts to La Niña the next year, then rainfall totals drop that following year, which is more common. The strong, 1982-83 El Niño, for example, brought 25 inches of rain to Los Angeles, but when the next year immediately swung to La Niña, LA got only 8 inches of rain. Same goes for the 1997-98 El Niño that brought 29 inches of rain to LA, but when La Niña came into play the next year, LA got only 9 inches of rain. This rapid shift from Niño to Niña (with no neutral in-between) can give a false perception that only El Niños bring heavy rain, which is not the case. Instead, when the less frequent, neutral Niño follows an El Niño the next year, rainfall can be equal to, or greater than the amounts that fell during the preceding El Niño winter. This happened not only in the neutral Niño of 1992-93, but also the Great Flood of 1861-62, the Christmas Flood of 1964, and more recently, the winter of 2016-17 — none were El Niño winters, but they weren't La Niña either. They were neutral Niño seasons; an El Niño occurred the year before; and they all saw atmospheric rivers, and subsequently very rainy winters.[76]

On top of all of that, there can be an even stronger influence on a neutral Niño rainy season from a unique atmospheric pattern that spins around the globe every 30-90 days called the Madden-Julian Oscillation, or MJO. This cycle not only influences the formation of an El Niño, but later it can pump up atmospheric rivers during the following neutral Niño year. Both cases happened recently, first in early 2015 (adding juice to the mega El Niño forming that year), and then in early 2017, enhancing atmospheric rivers during that wet, neutral Niño winter. I'll get to into the MJO further in the next section of this chapter to explain why it does what it does and how it plays into the bigger story here. But when you put it all together during a neutral Niño year — moderate jetstream, enhanced tropical moisture transport, and the MJO — you can have one heck of a rainy winter in Southern California, rainier than an El Niño winter. Add on top of this the effects of climate change, which have caused a more moisture-laden atmosphere from warmer than normal ocean waters (from additional evaporation), and it becomes clearer why the winter of 2016-17 was so wet.[77]

Atmospheric rivers though aren't always a bad thing. They are actually quite common for rain in California, the American West, and other parts of the globe. In fact, recent studies show that atmospheric rivers cause up to half of all extreme weather events at midlatitude regions (the area between the tropics and arctic regions). In Europe, almost 75% of the worst weather disasters were from wind damage linked to atmospheric rivers. And nearly half of California's total annual rainfall is from atmospheric rivers as well, yet it doesn't get the media's attention the way El Niño does.[78] Still, while atmospheric rivers can be a good thing, they can produce serious problems.

For California and the west coast of the U.S., atmospheric rivers usually don't last very long. But in the rainy season from October 2016 through March 2017 — when conditions were prime for atmospheric river formation and enhancement — 45 atmospheric rivers made landfall on the U.S. west coast. Out of those 45 events, 20 were measured as moderate but 12 were strong and 3 were rated as extreme.[79] A curious phenomenon helped this along, adding fuel to the atmospheric river storms bearing down on California.

The MJO

There's a curious weather pattern that circumnavigates the earth very 30-90 days that not only helped to strengthen the El Niño of 2015-16; it also helped to better establish the atmospheric rivers that brought flooding rains the following winter. I briefly mentioned this pattern — known as the Madden-Julian Oscillation (MJO) when talking about atmospheric rivers earlier — but having such an important influence on the events from 2015-18, it deserves a bit more attention. Here I'll give the MJO its due.

While El Niño comes around every few years (and can last for a year or two as well), and the PDO can come around every decade or more and linger for many years, the MJO is far more frequent; in fact, it occurs all the time, taking one to three months to circle the globe. You can think of the MJO as a boxing match between weather titans on an airplane in the upper atmosphere that constantly circles the earth near the tropics. When the MJO comes to your part of the planet, rainy weather and tropical cyclone formation becomes more likely as the MJO rumbles through. Once the MJO passes, things have a better chance of clearing up — as though the imaginary boxers rang the bell and went to rest in their respective corners of the ring.

The MJO has two effects that had a serious impact on the El Niño that formed in 2015-16, and then later when rains fell during the neutral Niño winter of 2016-17. First, as El Niño was gaining strength during the spring and summer of 2015 —

warming the equatorial waters, displacing sea life, disturbing upwelling cycles, and more — the MJO became quite active, which added to the fray of Niño-making ingredients. As warm waters became prime for typhoon formation during the summer of 2015, the stormy cycle of the MJO met up with that activity and provided an added punch to whip these storms into wilder frenzies. As these typhoons, starting in February 2015, blew hurricane force winds, they unleashed what are known as westerly wind bursts, or WWBs for short. Spinning counterclockwise, numerous typhoons in the northern portion of the Western Pacific blew bursts of days-long wind from the west to the east, which helps to move the warm, Western Pacific waters toward the El Niño zones in the Eastern Pacific. It's a way of speeding up El Niño's progress, making sure plenty of warm water is sent toward Peru and other points in the eastern equatorial Pacific. This is the first jab in the MJO's one-two-punch.

In late March 2015, a category 5 Super Typhoon named Maysak blew winds at 120 mph for days on end. Super Typhoon Noul did something similar in early May 2015, followed by yet another super typhoon, Dolphin, when Noul was about to lose strength. Around this time, before summer 2015 got underway, the MJO was quite active. And during the latter part of June 2015, the MJO became extremely active.[80] But starting back in March 2015, as the Pacific typhoon season got a hyperactive kickoff, exceptionally strong westerly wind bursts, influenced by the MJO and the typhoons it touched, were ongoing. This helped to move warm waters from the Western Pacific toward the El Niño zones in the east. It was like putting gasoline on a climate fire. El Niño wasn't just building; it was now getting steroid injections.[81]

But that's only the tip of the MJO's tail.

The MJO merely works its weather magic within the overlying climate cycles that wax and wane over longer, years-long episodes. When El Niño is afoot, the MJO can help it along, mostly in the spring of an El Niño year. When Niño though goes neutral — or swings to La Niña — the MJO behaves differently. In fact, once El Niño goes quiet, like it did during the rainy winter of

2016-17, the MJO can help establish atmospheric rivers that lead to torrential rains, and floods.[82] So after the MJO adds a punch to a strengthening El Niño, a neutral Niño season that follows gets a knock-out punch.

Imagining how the MJO may energize typhoons from its constantly stormy weather may be tough to visualize. Getting a grasp on how this same pattern could then arouse atmospheric rivers may seem even more puzzling. There are though just a few simple steps that make this happen, which unfolded in classic, textbook style during the winter of 2016-17, like so:

As the MJO travels around the globe from west to east, it eventually glides off the Indian Ocean where it kicks up heavy tropical rainfall. As the MJO continues on its eastward trek, it next enters the western tropical Pacific, where things start to get interesting. Almost immediately, a moisture plume starts to form ahead of the MJO's stormy activity in the Western Pacific, bringing with it additional moisture from the Indian Ocean, as well as the waters of the Pacific. This plume then stretches out over the next few days toward Hawaii, starting the formation of an atmospheric river, but at barely a trickle — at first.

As the MJO continues to cross the Pacific, moving east, during a non-Niño year it bumps up against high pressure sitting in the Gulf of Alaska — during an El Niño year that high wouldn't exist, but during a neutral Niño (or La Niña), high pressure tends to meander around the Gulf region. Once the MJO and the ensuing plume hit that Gulf high, although the main jetstream is guided north and over the high, a smaller jetstream forms to its south, which says hello to the approaching MJO (and the now-extended moisture plume). The formation of an atmospheric river pattern is now well underway with a long plume of tropical moisture extending from the eastern portion of the Gulf of Alaska to Hawaii. But there is one more element that is needed to ignite the rain event: low pressure, which is about to be called into action.

As the jetstream's winds blow around the northern area of high pressure in the Gulf of Alaska, and weaker winds (guiding the

MJO's stormy weather) blow south of the high, a trough of low pressure forms to the east of the Gulf high, which then spins off a stormy low pressure system, which then meanders off the west coast of the United States. Voilà! The last ingredient is thrown into the pot, and an atmospheric river storm begins to brew.

The MJO is a unique phenomenon that is mostly overlooked by TV forecasters and reporters. It's a complicated pattern that only increases probabilities, but is never a sure-fire thing. It can have multiple effects, which in our story here helped to initially fuel the mega Niño of 2015-16, and then enhance rainfall the following winter when Niño went quiet and the MJO assisted the formation of atmospheric rivers.

Although 2017 started out with a weak MJO, atmospheric rivers were starting to form by January 7, 2017, evident when a large but lumbering low pressure system setup shop off the Pacific Northwest. As it swirled counterclockwise, it created a southward bulge, known as a trough. But instead of extending directly south, this trough extended to the southwest, as though this storm were reaching out to Hawaii to grab all the moisture it possibly could. This started an atmospheric river pattern that brought about an inch of rain to Sacramento on the 7th, and then two inches a few days later. A few days before, after, and during this atmospheric river pattern, rain fell across much of the state for ten straight days spanning January 2nd through January 12th, thanks to an atmospheric river pattern — not El Niño.

Then came along the MJO, which became active by February 2017.[83] Atmospheric river patterns subsequently became more common that month as the MJO helped to build the much needed moisture plumes that slow moving low pressure systems, sitting off California, would tap into. When rainfall picked up late January into early February 2017, parts of California could take no more. An overabundance of rain had fallen from these atmospheric river storms, and a new kind of crisis was about to unfold.

Oroville by Example

Everything came together to make one heck of a rainy season around California in the winter of 2016-17. Lake levels were on the rise, especially by early 2017, as a series of Pacific storms kicked off the new year with rain starting January 2nd. A more notable series of three potent storms brought another round of heavy rain for nearly six straight days from January 18th through the 23rd. These storms, while bringing a heavy amount of rain to various parts of California, also kicked up high surf that reached Southern California with 12- to 15-foot breaking waves. Tragically, two women were swept out to sea in San Diego from one particular swell — only one survived. Snow also fell in the mountains as far south as Southern California measuring 2-5 feet as low as 4,000 feet in elevation. Precipitation was coming in from all angles, and Californians rejoiced — for a while.

The series of exceptionally wet, early January 2017 storms were tapping into colder, Artic air (hence the snow), and the moisture they tapped into was mostly from their weather fronts and somewhat disorganized plumes of moisture ahead of them. That's how it started, but it's not how it ended. Things would change shortly. Atmospheric river patterns were barely starting to get underway in January 2017 and they wouldn't be as well formed until February, when the California rain machine kicked into high gear. Still, with three, back-to-back storms during the second half of January, California's reservoirs were well on their way to being filled. More rain, though, was on the way — this was just a warmup for the main event.[84]

By February 5, 2017, a steady flow of tropical moisture had formed a substantial, unbroken moisture plume that extended from Northern California to Hawaii. The MJO was doing its thing as it entered the mix, helping to extend that moisture plume. Once the MJO met up with a high pressure system parked in the Gulf of Alaska, it further assisted this long plume of tropical moisture as a slower moving jetstream formed below the Gulf high. It was a textbook, MJO-influenced atmospheric river. In true MJO form, a low then spun off the east side of what would have otherwise been a storm-blocking high pressure system in the Gulf, which then formed a perfect blend of plume and storm to stir in a lengthy period of precipitation to California. The Pineapple Express — as this particular kind of atmospheric river is known — was here. But the MJO was just starting to have a positive influence on atmospheric river formation, and it would have a more profound effect in the coming weeks. On February 5, 2017 though, rain started to fall across California — and it wouldn't stop any time soon.

Lake Shasta in Northern California rose from 900 billion gallons of water in February 2016 to 1.2 trillion gallons by February 2017, even higher by March.[85] This changed the water elevation in Lake Shasta by more than 100 feet, from a low of 914 feet near the end of 2015 to an eventual 1,048 feet by the end of March 2016.[86] Shasta, like many lakes in California, reached or exceeded capacity. There was plenty of water to go around to water the "little green lawns" Governor Brown talked about a couple of years earlier near the end of the drought. And when the storms were all said and done, drought restrictions in California were lifted by April 2017 — with much less fanfare from Governor Brown and the media.[87] But when it came to the rain earlier in the year, like so many other things in life, too much of a good thing can turn bad.

About 100 miles to the south of Shasta, Lake Oroville — home to the tallest dam in the United States — exceeded its water storage capacity by February 2017, before the newly forming atmospheric rivers developed and commenced. The rain from January was excessive enough to strain this lake's capacity. Rising from a low of 600 billion gallons in February 2016 to nearly 900 billion gallons by February 2017, Lake Oroville's spillway became overly stressed, with a serious crisis now looming. All the while, atmospheric storms lay on the horizon.

Inflows into Lake Oroville were especially rapid at the end of January into the beginning of February 2017; in fact, on one day alone, February 6, 2017, as rain fell across California from a newly formed atmospheric river extending a continuous feed of precipitable water from Hawaii to California, inflow to Lake Oroville increased from 30,000 cubic feet of water per second to over 130,000 cubic feet per second — more than four times the pace and volume in just a single day. By February 9th, after an extended period of rain, inflow rate increased even further to over 190,000 cubic feet per second As water rushed down the main, concrete spillway at Oroville, a crack appeared on February 7th. But rain wouldn't stop for another three days.[88]

To get a handle on the situation, water was abruptly stopped from flowing further down the main Oroville spillway, allowing inspectors a chance to see what was going on. As water backed up in the lake, the inspectors noticed a large, 250-foot wide crater about halfway down the concrete spillway, spelling bad news, to say the least. While the dam was holding, the eroding spillway was signaling an alarm that the dam, too, may give way.

Officials opened the emergency spillway at the Oroville dam, yet this was done with trepidation. Community groups, concerned of a potential breach, requested $100 million the year before to upgrade the emergency spillway since it had become vulnerable to erosion. Their requests though fell on deaf ears, and the Oroville upgrade was denied by federal regulators. Ironically, federal regulators did an about-face later in 2017 and asked state officials to explain why there was new damage to the Oroville spillway years after this whole thing was resolved — that though is a story for another time.[89] I digress...

With the main spillway compromised and crumbling in February 2017, Oroville dam officials had little choice but to reroute water through the sketchy emergency spillway until repairs could be made. Nobody knew though what could happen next, but it was a risk that had to be taken. Evacuations were ordered for the low-lying areas in the path of the Feather River Basin, where spillway waters flow. But, this didn't happen until a day *after* the emergency spillway was opened. As dam officials worked to resolve the issue, entire towns lay downstream from a potentially fatal disaster.

Knowing that the emergency spillway was in questionable condition — and was never used before — while realizing that releasing too much water could unearth trees and surrounding land through the emergency, earthen spillway, officials released, at first, a relatively small amount of water at a fairly safe rate of 20,000 cubic feet per second. That almost immediately widened the emergency spillway's hole by another 50 feet — that effort was stopped. With both spillways now shut off, water continued to fill Lake Oroville until it overflowed over the lip of the dam, down the emergency spillway; there was no place left for it to go. Extensive death and disaster were now real possibilities for 180,000 people in the path of potential flood waters.

As the dam overflowed, it eroded the hillside along the emergency spillway. Officials saw that this could undermine the entire dam, which could then release a 30-foot wall of water into the communities below. Evacuation orders were finally given, and as people were preparing to flee, dam officials attempted to ease the dangerous erosion undermining the dam by releasing even more water, which as you might expect, worsened the problem, damaging the emergency spillway further. Luckily though, the dam held.[90] Had it not, over 180,000 people would have been in the path of unstoppable flood waters, and would have likely perished.

Oroville though wasn't the only flood crisis in California in 2017 as the rainy season that year brought precipitation to the entire state. The Russian River in Sonoma rose three feet above flood stage in 2017, inundating close to 500 homes, causing 570,000 residents to lose gas and electricity, forcing 3,000 people to evacuate, and causing millions of dollars in crop damage. In the Silicon Valley's city of San Jose, the Anderson Dam overflowed, forcing the evacuation of 14,000 people and causing $73 million in damage. The San Joaquin River, running through Central California's agricultural heartland, suffered a levee breach, causing 500 people to evacuate. The Big Sur highway was shut down at times from multiple mudslides and a collapsed bridge. Southern California dealt with sinkholes, and at one point Interstate 5 in Sun Valley was inundated by more than two feet of water, trapping motorists in their cars.[91]

The flooding around Oroville and other parts of Northern California were not just from one atmospheric river, but multiple such rainy events over a span of a few months. This book's cover, by the way, shows a NASA image of one such, hefty atmospheric river that was well entrenched along the California coast on February 20, 2017. The storm that swirled in this atmospheric river started a few days earlier, affecting California in a variety of ways.

On February 16, 2017, four days before the NASA image on this book's cover was shot, a strong area of low pressure sagged south from the Gulf of Alaska, parking itself off the coast of Northern California, tapping into the atmospheric river plume that was just starting to form into a constant stream of moisture, which would bring a continuous river of rain to the region for the next five days. This low pressure system, sitting and spinning off the coast of California, was strong enough to whip up strong winds and high seas, that brought dangerous surf to the entire California coast. In fact, by the afternoon of February 18th, Oceanside Beach lifeguards reported waves near 16 feet — about 400 miles south of the storm. Sitting so close to the coast, this storm also brought strong southerly winds from its counterclockwise, cyclonic spin, which downed trees at the coasts and valleys. Mountain ski resorts got an additional couple feet of snow.[92]

Fortunately for Oroville and the residents in the path of its potential flood water discharge, rain that continued to fill the reservoir flowed safely through the emergency spillway. Disaster was averted, even as rain continued to fall. But Oroville got lucky — this time.

In January 2018, final word finally came in on the Oroville crisis from an independent forensic team report analyzing what went wrong.[93] Mulling over mounds of paperwork, conducting research, analysis, and interviews, the report found that there wasn't just one single thing that went wrong; everything did. Physical factors included vulnerability in the spillway's design and construction, including poor spillway foundation conditions in some locations. Cracks since the 1970s were thought to be safe, and normal. And the report also pointed out that the decision to use the emergency spillway against the advice of civil engineers was a mistake. In conclusion, the Oroville forensic report ended with a chilling summation:

> *The fact that this incident happened to the owner of the tallest dam in the United States, under regulation of a federal agency, with repeated evaluation by reputable outside consultants, in a state with a leading dam safety regulatory program, is a wake-up call for everyone involved in dam safety.*

Oroville became the poster-child of modern-day California flood disasters, setting an example of how flooding rains during a neutral Niño year could threaten the entire state. Yet the story of Oroville has, for the most part, a happy ending, as do most accounts of the flooding rains in 2016-17 for California — at least when compared to the Great Flood of 1861-62. But not all stories of heavy rain end happily ever after; atmospheric river storms hit more than just California in 2017; and not everyone sitting in the path of a dam's spillway would be as lucky as those near Oroville.

Harvey's Alarms

California is not the only portion of the United States prone to extreme rain events from atmospheric rivers. The Gulf coast, from Texas to Florida, sits in prime territory for these kinds of flooding events as well. After atmospheric rivers brought flooding rains to California in the first few months of 2017, another atmospheric river would form near Texas late in the summer as El Niño remained in a neutral state, ideal for atmospheric river formation. The atmospheric river that summer would be amplified by a powerful hurricane, and together they would unleash a watery hell in August 2017, tying Katrina as the costliest hurricane to ever hit the United States. Hurricane Harvey, which set numerous records, served as a hallmark for not just hurricane damage, but also atmospheric river devastation.

Although the Pacific has its Pineapple Express — an atmospheric river that sometimes stretches from Hawaii to the U.S. west coast — the Gulf Coast of the United States has the lesser known Maya Express, which helped fuel Harvey.[94] Like most hurricanes, Harvey formed off Africa and trekked along tropical latitudes until it ended up in the Gulf of Mexico, where it was further fueled by the extremely warm waters in that region. Things weren't looking all too bad though until Harvey reached the Bay of Campeche, just east of the Yucatan Peninsula. But it wasn't just the warm Mexican waters that helped feed Harvey into a mega hurricane; it was the neutral Niño that turned things around.

While the Maya Express, like most atmospheric river formations, can spring into action during a neutral Niño, there was also a lack of El Niño-driven, jetstream winds to stop this atmospheric river's formation and guide it as well. As you may recall from earlier discussions, during an El Niño, as the Pacific jetstream drops toward SoCal latitudes, it also lines up over hurricane alley, where Harvey was gaining strength. If the jetstream and its strong winds get draped over that region then it causes wind shear, which blows the tops off of rising hurricane columns, putting the kibosh on hurricane formation. But that wasn't the case in 2017; instead, being a neutral Niño year, the jetstream was weaker and not placed over hurricane alley. This allowed storms like Harvey to grow into humongous hurricanes. But as Harvey grew, it reflected a common misconception and contradiction regarding hurricanes and climate; while at the same time, an ex-VP was hawking his brand new climate wares, and ringing the alarm. Al Gore found Harvey to be a great source for free advertisement.

Although later redacted in recent IPCC reports, there was a time when climate studies called for more frequent El Niños and Katrina-like storms. Those assumptions have since been changed by the IPCC — but not Al Gore, who, as Harvey and other hurricanes spun-up in 2017, once again came at odds with the experts, as well as the science behind that year's spat of disastrous hurricanes. Harvey was serendipitous for Gore in that, while Harvey was forming, he just so happened to be on a publicity tour for his new sequel, called simply, *An Inconvenient Sequel.* Gore took this opportunity to once again exploit climate change, warning Texans that global warming will lead to more hurricanes like Harvey in the coming years.[95] This rang the first of two alarms regarding the wrath of Harvey. The second alarm — to get out of Harvey's way — went sadly silent for way too many, making a bad situation even worse. While the latter caused an immediate crisis that I'll get to shortly, it's worthwhile to pull off the Harvey-highway for a second, and take look at the former — since it almost stole the spotlight from the actual problem at hand, based on false assumptions and outdated information preached by the gospel of Gore.

As soon as Harvey formed into an impressive hurricane in the Gulf of Mexico, Katrina and climate came to mind, but in a conflicting way. A big El Niño had recently occurred, but we know that more frequent El Niños would not result in more Katrina-like storms like Harvey. If El Niño were to blame, then it would have blown Harvey to bits before it could form (from the Niño-guided jetstream causing wind shear). And of course, Harvey was forming during a neutral Niño year, so that idea (linking El Niño frequencies to hurricanes and climate change) was out of the question. But could climate change be responsible in some other way for Harvey growing into a massive storm?

We have to remember that much of what is talked about today regarding climate change is taken straight off the pages of Al Gore's *An Inconvenient Truth*, which is based on a report by the IPCC from 2001. More recent reports by the IPCC tell a different story, but many (including Al Gore) continue to base their assumptions on the older report, not the newer ones (there have been two IPCC reports since Gore released his original book). These newer IPCC reports have refined — and now disagree with — earlier assumptions regarding links between climate change, El Niños, and hurricanes like Harvey. Even though the Maya Express was a primary contributor to Harvey during a neutral Niño year, it's not what Gore and other alarmists gave attribution to; instead, they continued to refer to outdated information while ringing the wrong alarm. The IPCC's most recent assessment report from 2013 now states that while El Niño would remain the dominant means of weather variation across the tropical Pacific for years to come, there is now "low confidence" that El Niño will vary all that much from climate change. In fact, the IPCC has removed all wording of any increase in El Niño frequency from its most recent reports.[96]

Hurricanes, like Harvey, become questionable in those more recent IPCC climate reports. While the IPCC's report in 2001 (known as the TAR) called for an increase in tropical cyclone frequency and intensity, the latest IPCC report (the AR5 from 2013) downplays that, stating there is "low confidence" that any changes in tropical cyclone activity are linked to any kind of human influence, and that natural causes, instead, play a major role.[97] To further muddy the waters on what to expect from hurricanes like Harvey, the IPCC's latest report states:

> *...it is likely that the global frequency of occurrence of tropical cyclones will either* ***decrease or remain essentially unchanged****, concurrent with a likely increase in both global mean tropical cyclone maximum wind speed and precipitation rates. The future influence of climate change on tropical cyclones is likely to vary by region, but the specific characteristics of the changes are not yet well quantified and* ***there is low confidence in region-specific projections of frequency and intensity****.*

(Bold added by me to denote pertinent wording.)

In other words, when it comes to hurricanes in a warming world, the IPCC now believes that there will either be fewer hurricanes going forward, or the number of them will stay the same — not increase as earlier reports had stated. All the while, the IPCC says there is *some* possibility that hurricanes could become more intense in the future, but there are too many conflicting studies to support that for sure. So while the finger of blame for Harvey was being pointed toward climate change by Gore and others in 2017, that was an incorrect assumption based on obsolete reports. Although science can change; dogma never wants to. Science though, is what Harvey was all about.

Not having a direct link to climate change does not make Harvey any less of a concern. There is no sigh of relief. Instead, if nature could do this all on her own, then we should be even more concerned, since it means we're unprepared for all the inner workings of the natural world. Removing a climate link to El Niño and hurricanes means we now need to shift our attention to a potentially more serious issue.

When it came to Harvey — and where the rubber really meets the road — it wasn't El Niño or a link to climate change that made this storm the monster that it grew into; instead it was the absence of any major jetstream pattern over hurricane alley during a neutral Niño year, which influenced the formation of a major atmospheric river storm that is historic enough to be given a name — the Maya Express. It doesn't make for catchy headlines the way global warming does, but sometimes the truth in science is not so inconvenient, but just mundane. What Harvey *did* though was anything but.

As Harvey sucked up warm Gulf waters off Mexico, that storm's counterclockwise rotation began to tap into the Maya Express's plume of moisture extending south across the Yucatan Peninsula, across Central America, and into the Pacific. As Harvey sat there, spun, and pumped up in strength, it then moved north, making a beeline for Texas.

On August 25, 2017, Harvey made landfall near Rockport Texas as an intensely strong Category 4 storm packing 130 mph winds — almost twice as strong as what was forecast by the National Hurricane Center earlier in the day. Harvey was also moving along at a glacial pace of only six knots, and by Saturday morning on the 26th Harvey slowed to a near stall at just two knots — an incredibly slow speed for a hurricane, putting it at nearly a standstill. As Harvey slowly sat and spun off the coast of Texas, he not only had more time to suck up copious moisture, but moving so slowly, Harvey was given a longer length of time to spit it back out once reaching land — a classic trait of an atmospheric river storm. With slow-moving Harvey poised to bring long periods of precipitation, warnings from the National Hurricane Center were sent out, calling for exacerbated rain with a high likelihood for catastrophic flooding — from rainfall that could last for another week.[98] But another unwanted surprise was in store.

On August 28th, Harvey drifted back into the Gulf of Mexico, stalled again, and then eventually retrograded back toward the coast for another round of punishing rain. Harvey later picked up a little momentum and moved away from land, back into the Gulf. This wasn't the end though; it was a refueling mission. On August 29th Harvey made landfall once again in Louisiana and didn't dissipate until two days later. But more rain was falling in east Texas.

Water rapidly filled lakes and reservoirs around Houston to the point of near collapse. On August 29th, a levee breeched along the Columbia Lakes, located south of Houston. Evacuation orders went out for the surrounding area but luckily the levee was fortified in time to stem the flow of flood waters. This wasn't the case though for the Addicks and Barker reservoirs sitting just east of Houston. In similar fashion to California's Oroville crisis earlier in 2017, the dams at the Addicks and Barker reservoirs were nearing a state of compromise as water started spilling out over top. This forced the hand of the Army Corps of Engineers to release a torrent of flood water into the Buffalo Bayou that runs through residential areas near downtown Houston. Water rushed into thousands of homes that would have otherwise been protected by these dams.[99]

When it was all said and done, Harvey had dumped an unprecedented amount of rain from Corpus Christi to Houston and even Beaumont, Texas farther east. All told, more than 60 inches of rain fell over southeast Texas, setting the record for the highest precipitation total observed from a tropical cyclone in the United States.[100] Harvey caused an estimated $125 billion in damage, and 91 people lost their lives. Luckily, two days before Harvey hit Texas, Governor Greg Abbott declared a state of emergency for 30 counties. Mandatory evacuation orders were issued for more than half a dozen counties. But strangely enough, in Houston, a city known for its susceptibility to flooding, evacuation orders were not given — a flashback to Katrina reappeared.

Houston's mayor, Sylvester Turner, never gave the signal to get out, leaving Houston residents with a false sense of safety as floodwaters rose. Turner, when questioned why he didn't warn the people of his fair city to get out of harm's way, attempted to justify his non-committal by saying,

> *You literally cannot put 6.5 million people on the road. If you think the situation right now is bad, you give an order to evacuate, you are creating a nightmare.*[101]

Despite evacuation and emergency management plans enacted in all major cities across the United States, Mayor Turner, for some reason, didn't put faith in his own system, or the people he was there to protect and serve. Instead, he created unnecessary chaos and catastrophe.

Thousands of Houston's residents were left stranded, many on their rooftops, while others waded through flooded streets with children riding on adults' shoulders. Yet Mayor Turner insisted that it was best for people to stay in their homes, as water levels rose. Sadly, the Texas Department of Public Safety reported that more than 185,000 homes were damaged in Texas from Harvey, and 9,000 homes were completely destroyed. Approximately 32,000 people were displaced in shelters across the state, and more than 210,000 people registered with FEMA for disaster assistance.

We'll never know if Turner made the right call, since things may have been worse if a mass exodus had commenced. But we do know that rescue and recovery efforts started much quicker than they did during the Katrina crisis; if they hadn't, the results could have been much different. What we also know, is that Harvey was just one devastating storm in 2017 — there were more on the way, and it was about to get ugly.

The Irma Conflict

Nipping on Harvey's heels was another hurricane that set the Gulf and East Coast states on high alert. Hurricane Irma set off on the typical path from the tropical eastern Atlantic, gaining strength as it headed toward the Caribbean islands. After gaining Category 5 status, Irma either touched or punched nearly all of the islands in the area, affecting Antigua, Barbuda, the British Virgin Islands, Puerto Rico, Turks and Caicos, Puerto Rico, and Cuba, among others, before setting Florida in its crosshairs.

Instead of drifting into the Gulf of Mexico like Harvey did, Irma, once passing Cuba, turned north and blasted the Florida Keys. Models had originally forecast Irma to plow through the center of Florida as a Category 5 storm, which would have destroyed much of the state. However, Irma weakened slightly to a Category 4 storm once hitting the keys, and then weakened a bit more before traversing north through mainland Florida.

To list the destruction would take volumes beyond this book. But at the end of the day, Irma caused more than $65 billion in damage, and was responsible for 134 deaths.[102] Surely something was amiss, with another powerful hurricane wreaking havoc in the Atlantic. But since El Niño wasn't around in 2017, blame shifted to the next best thing in reach: climate change.

It was clear once the 2017 hurricane season kicked into high gear that this was not going to be a normal season. People in the potential paths of hurricanes were on edge to say the least. But the tell-tale signs that El Niño had gone quiet were more than evident as Atlantic hurricanes seemed to be turning into leviathans at the drop of a hat. All the elements had come together as El Niño waned: Waters were warm…check. The atmosphere was laden with moisture...check. Wind shear was minimal...check. Islands were located in hurricane alley...well...check?

The Caribbean is constantly dodging hurricane bullets as they get side-effects year after year from powerful storms that often just brush by them. No climate shift or cycle will change the fact that these islands are sitting ducks — very *lucky* ducks most of the time — sitting in the line of fire, just waiting to be slammed by tropical storms. And in 2017 there were two notable hurricanes (Irma and Maria) that, instead of taking a less invasive course *around* Caribbean islands, changed course ever-so-slightly and went through them. The climate bell was rung on cue, this time by people in high places.

Despite reports by the IPCC — which is funded by the United Nations — the UN Secretary-General, Antonio Guterres said at a high-level event on Irma:

> *The season fits a pattern: changes to our climate are making extreme weather events more severe and frequent, pushing communities into a vicious cycle of shock and recovery...Reducing carbon emissions must clearly be part of our response, together with adaptation measures. We must be able to bend the emissions curve by 2020. The rise in the surface temperature of the ocean has had an impact on weather patterns; and we must do everything possible to bring it down.*

I'm sure that Mr. Guterres speaks with the purest of intentions, but this is an ironic case where the right hand (the head of the UN) doesn't know what the left hand (the UN-funded IPCC) is doing. Instead, Guterres was taking his climate cues from an ex-politician (Al Gore), not the scientists his organization funds. Yes, ocean waters have warmed and that is linked to human-induced climate change. But, as pointed out earlier, the IPCC has pulled back on linking climate change to hurricanes. Additionally, NOAA — which oversees the National Hurricane Center — in a report many months after Irma, continued to urge caution when trying to connect hurricane activity to climate change.[103] To be clear, and not cherry-pick bits and pieces to support any confirmation bias, NOAA says, in full, that:

It is premature to conclude that human activities — and particularly greenhouse gas emissions that cause global warming — have already had a detectable impact on Atlantic hurricane or global tropical cyclone activity. That said, human activities may have already caused changes that are not yet detectable due to the small magnitude of the changes or observational limitations, or are not yet confidently modeled.

Despite the lengthy twelve years between the time of the IPCC's third assessment to their recent AR5 report, no one has successfully been able to model a link between human-induced climate change and hurricane activity. Twelve years, and still no causal link. To be clear, human-induced climate change is real, but it can't be linked to every single weather event; it's something the weather experts agree on, especially when it comes to hurricanes like Harvey and Irma.

According to NOAA's statement, the worst-case scenario to date is that there may be minor (small magnitude) changes that we cannot yet detect (despite more than a decade of research, during which time many more things were successfully linked, but not hurricanes). So as it stands today, hurricanes like Irma are not being linked to human-induced climate change; however, there may be things we're doing now that could change that link in the future. This becomes a bit clearer from NOAA's report on "Global Warming and Hurricanes" when it says:

There are better than even odds that anthropogenic warming over the next century will lead to an increase in the occurrence of very intense tropical cyclones globally — an increase that would be substantially larger in percentage terms than the 2-11% increase in the average storm intensity. This increase in intense storm occurrence is projected despite a likely decrease (or little change) in the global numbers of all tropical cyclones. ***However, there is at present only low***

***confidence that such an increase in very intense storms will occur in the Atlantic basin*.**

(Bold added by me to denote pertinent wording.)

In other words, putting this all together, NOAA, the agency that oversees the National Weather Service and the National Hurricane Center, has determined there is no link between present-day hurricanes and human-induced climate change; furthermore, NOAA feels that while the future could change and hurricanes could become more powerful from climate change, it would not likely affect the Atlantic, where Harvey and Irma brought unthinkable loss of life and property. But with the general population hearing climate blame from the media, Al Gore and now the head of the United Nations, Irma was being directly linked to the hand of humankind. It was another unnecessary distraction — lowering your carbon footprint was not going to affect hurricanes like Harvey or Irma, but concentrating efforts instead on mitigating these kinds of *natural* disasters would be far more prudent.

Nevertheless, while Irma was given an inaccurate attribution, in many ways it was just our innate human nature. Similar to our ancient ancestors who built Stonehenge, sacrificed animals or people, and performed ritual dances in an effort to change the weather by admitting their sinful deeds, when it comes to climate change we too have a tendency to self-flagellate, thinking we are the cause for hurricane harm; thus, directing attention away from the real, immediate problem at hand. This would play out on the world stage when another hurricane in 2017 would cause devastation on such a grand scale that it pulled back the curtain of not necessarily climate, but lackadaisical preparation, incompetence, and government corruption.

The Maria Test

On May 25, 2017 forecasters at NOAA's Climate Prediction Center predicted that the summer and fall would see above-normal hurricane activity. They were right. For the Atlantic hurricane season, which runs from June through November, NOAA forecasters predicted a high likelihood for 11-17 named storms, of which 5-9 could become hurricanes; and of those, 2-4 could be major events. Living up to the high end of those expectations, the Atlantic hurricane season ended with a total of 17 named storms, 10 of which were hurricanes, and 6 of those were major — a tad higher than the forecasts made earlier in the year. But not only was 2017 a hyperactive year for hurricanes, it was also a time of extreme events.

A couple weeks after Harvey calmed down, a new system was spinning across the tropical Atlantic, setting its sights on Dominica, and, more memorably, Puerto Rico. Hurricane Maria would cut a swath of destruction when it passed through Dominica on September 18, 2017 as a powerful Category 5 storm, packing 175 mph winds. After losing some steam once passing over Dominica's small land mass, Maria held onto Category 4 status as it plowed smackdab through the center of Puerto Rico. No Caribbean island was completely sheltered from Maria as this storm impacted all of the Lesser Antilles — not just Dominica but the U.S. Virgin Islands as well.

Once Maria finally turned north and later fizzled far out in the Atlantic, the damage was tallied at almost $100 billion, and more than 500 people had been killed. The entire power grid of Puerto Rico was effectively destroyed, leaving the millions who lived on the island without power; in fact, one month later, nearly 90% of the island would *still* be without power and only half of the sewage treatment plants would be operational. Three months after Maria, almost half of the island continued to be in the dark. Drinking water became unsafe as an outbreak of leptospirosis — a severe bacterial infection caused from animal urine — affected residents stranded on the island. Maria didn't just punch Puerto Rico; it pummeled the island to the ground, and it would not recover any time soon. But like two fighters in the ring, it's not necessarily the hit, but who (or what) got hit.

Puerto Rico suffered a different kind of disaster than Houston did, evident by a stark contrast in destruction compared to the storms that caused it. Harvey dumped 60 inches of rain on Houston while Maria brought about half that much to Puerto Rico (38 inches of rain). Both storms made landfall as Category 4 storms, and Harvey caused 37% more in damages than Maria did. But Houston was back on its feet relatively quickly, while Puerto Rico took much longer to recover. Even though Houston's record rain exceeded that of Maria — not to mentioned the additional release of reservoirs in east Texas to add to the mess — the recovery efforts between these two disasters varied widely. The reasons why are dissimilar as well.

Maria echoed Katrina's swansong as help *seemed* to come slowly and an unproductive political blame game got underway. The mayor of San Juan, Carmen Yulin, blamed President Trump, and Trump blamed the mayor. After a cataclysmic event like Maria, no one wants to readily own a potentially mismanaged, months-long relief effort, and "He said, she said" arguments are too often riddled with such an overabundance of bias that separating truth from dogma lets Rome burn while the arguing parties fiddle in flooding waters. But the fact remains that there were a number of obstacles to overcome in Puerto Rico that didn't exist in Houston; namely, infrastructure.

In 1973, when I was ten, I spent the better part of that summer in Puerto Rico when my father was temporarily sent there for work. While we stayed in the suburbs of "new" San Juan, I can still recall hillsides lining the greater part of the outlying areas dotted with an endless array of ramshackle slums — homes hobbled together with plywood and corrugated metal sheets covering dirt floors. Wild packs of dogs roamed the streets, even near our home in the better part of town. And I still remember to this day a chance encounter with a couple of boys my age, walking a tattered, sickly looking horse — which I found fascinating and drew me over to talk with these boys — through the parking lot of our church in San Juan where church-goers were getting into cars to drive home after service. With me knowing less than ten words of Spanish and those Puerto Rican boys likely knowing less English, none of us could understand each other when we spoke. But as flies swarmed around their horse, led not by a bridle but by a piece of old rope, and seeing the dirty clothes these boys wore with no socks to protect their feet from shoes that didn't fit them, I knew that their outstretched hands meant they were hungry. The encounter was brief, but unforgettable. Although our family, supported by my dad's meager salary as a minister, never had much money, that brief encounter with those Puerto Rican boys opened my eyes to the plight of others, and made me thankful for what we had — and what I still have today. It's something that never left me, and it's something I recalled while watching the news about Hurricane Maria.

When I think of Maria hitting Puerto Rico I think of the people that live in the slums and other poor neighborhoods around the island. Those boys I met, like me, are old(er) men now — if they survived the harsh life that laid before them. The slums in which they lived were barely habitable. Even in our relatively richer neighborhood in San Juan — what would be considered middle class, at best, in the U.S. — power and water were unreliable in 1973 due to sabotage and strikes from unhappy union workers, and unstable utility companies as well. If you've ever vacationed in the Caribbean and left the comfortable confines of your resort, I'm sure you can attest to the below-standard construction of roads, buildings, and the electrical grid, which in places resembles spaghetti. But the people aren't to blame.

The Puerto Rican government — which has a reputation for shady, banana-republic-like leadership — mismanaged their money so poorly that just a few months before Maria hit the island they declared bankruptcy. In their court filing in May 2017, the government of Puerto Rico stated that they were "unable to provide its citizens effective services" because of their mounting debt, including $74 billion in bond debt, and $49 billion in unfunded pension obligations.[104] With Puerto Rico asking for $94 billion out of the $100 billion in total damages caused by Maria, it's a lot like a run-down house getting blown to bits, then asking the insurance company to remodel it when they rebuild. This sentiment was echoed by Jorge Rodriquez — the Harvard-educated CEO of PACIV, an international engineering firm based in Puerto Rico — in an editorial in the New York Post a couple of weeks after Maria exited Puerto Rico. Rodriquez went on a controlled rant stating:[105]

> *For the last 30 years, the Puerto Rican government has been completely inept at handling regular societal needs, so I just don't see it functioning in a crisis like this one. Even before the hurricane hit, water and power systems were already broken. And our $118 billion debt crisis is a result of government corruption and mismanagement.*

Rodriquez went on to criticize Puerto Rico's governor, Ricardo Rossello, pointing out that he was only 36, never really held a job, never dealt with a budget, and hired an inexperienced administration. Rodriquez praised the FEMA efforts while pointing out embarrassing mistakes made by the Puerto Rican government which, when things didn't go well, didn't blame themselves but instead turned on the U.S. federal government. It's unfortunate that the Puerto Rican government sat on their laurels too long, getting far too lucky year after year not to be hit with a devastating storm. While Puerto Rico sits in the sights of approaching tropical storms, it's not at all funded or built to withstand the impacts of hurricanes when they come — and they will.

Being in the center of Hurricane Alley puts all of the Caribbean in peril, yet most of the population — and the governments that often squander their tax dollars — do not have the same economic means to build cities safe enough for ravaging, Category 4 hurricanes like Maria. When a hurricane like Harvey slams the United States, buildings are damaged, streets flood, and people die. But the wealth of our nation, along with better planning and building codes, provides a response and recovery far more rapid than the less fortunate living on many islands in the

Caribbean, like Puerto Rico. Living on a small island in low economic conditions, when a storm ravages your homeland there is just simply nowhere to go. Most of the people of Puerto Rico had to stay put, ride it out, and try to rebuild what they could.

FEMA reported that 60,000 homes needed roofing in Puerto Rico, but they had only 38,000 tarps to distribute. They also provided more than 6 million gallons of bottled- and bulk-water, but that was less than 10% of the drinking water needed for the entire territory, let along water needed for hygiene and cooking. FEMA provided water purification tablets and numerous mobile water filtration systems, but with communications so disrupted across the island, it was difficult to spread word to the people in need. As a result, some people resorted to federal hazardous waste sites searching for anything to drink. It's easy to point blame under these kinds of conditions, but providing aid to Puerto Rico is a far cry different from Houston.

FEMA's resources were stretched thin by the time Maria hit Puerto Rico since 2,600 of FEMA's staff were still busy with recovery efforts from Harvey, and others were still dealing with destruction from Irma. FEMA though was able to deploy 20,000 staff to Puerto Rico for the Maria effort, but when dealing with an island having a population of more than 3.4 million people, the 20,000 relief workers were just a drop in the bucket.[106] A number of U.S. Coast Guard and Navy vessels arrived on scene as well, but the uphill battle to bring Puerto Rico back from the brink of collapse would take time. As of February 2018, nearly five months after Maria ravaged Puerto Rico, recovery efforts remained ongoing, and obviously-inept Puerto Rican officials continued to denounced the federal government's response. More than 400,000 residents continued to be without power at the time of this writing.[107]

All too common, when a powerful hurricane rips through a region, the first thought is often climate change, but that line of thinking is unproductive. Despite knowing from reports by the IPCC and NOAA that these hurricanes aren't linked to climate change, complaining about not being green does little good. Driving a Prius or using curly fluorescent lightbulbs won't help the people of Puerto Rico today, nor will it tomorrow when the next hurricane will inevitably hit that island — or others nearby. Instead of pointing the finger of blame on something experts agree is not the issue, we should instead be looking for ways to ensure that when storms *do* arrive — and more will — people are prepared, and can recover more quickly. This is not to say that we should throw climate caution to the wind — quite the contrary. While there are some things we can do to mitigate human-induced climate change, there are more immediate needs of higher costs and consequences that should first be addressed.

Maria eventually passed, but her effects remained many months later. If Maria had not made landfall, there wouldn't be much to talk about; it would just be a short-lived headline and eventually a series of archived weather reports. Maria though *did* make landfall, which turned her into a rather salient system — a reminder that these nasty storms don't always stay out to sea. Yet between the time of Irma and Maria, a nearly equally powerful hurricane swept through Hurricane Alley: Jose. Staying out to sea, as many Atlantic hurricanes do, this powerful yet mostly unmentioned system fell through the cracks and was barely noticed — unless, that is, you surf the east coast of the U.S.

Between the Cracks

From September 6th through the 11th in 2017, social- and traditional-media exploded with viewer traffic searching for news on Hurricane Irma. But by the time Maria made the news about ten days later, hurricane interest seemed to drop off the map. Google Trends — a way to show statistical interest in something — shows that interest in Maria reached only 25% of what Irma did, and for a much shorter time.[108] Hurricane Harvey, by comparison, received about 35% interest compared to Irma. Even though Irma resulted in the fewest deaths and lowest cost in damages from these three storms, Irma seemed to be of the most interest when hurricane season was nearing its peak in the Atlantic in 2017.[109] More than likely this is from an initial wakeup-call once Harvey flooded

Texas, giving the next hurricane, Irma, the spotlight. Irma was also headed into Florida, with forecasts showing it could impact most of the state. But by the time Maria moved through Puerto Rico, everyone was already aware of hurricane damage that year, so it's likely that interest just simply dropped, especially for those not connected to the Caribbean, or Puerto Rico in particular. Distant disasters are more difficult to empathize with, and after a while any flood of news can make one immune to similar stories that occur around the same time. Such was the case for a couple of other interesting developments that fell through the cracks between Irma and Maria.

On September 8, 2017, a NASA satellite took a stunning picture of three, simultaneous hurricanes barreling through the Atlantic: Irma, and the lesser talked about Katia and Jose.[110] Katia's lack of media attention is justified as it only reached Category 2 status and had much lower impacts than Irma or Harvey. But the third storm was a different story.

Hurricane Jose got much less attention from the time it earned cyclone status on September 5, 2017 until it died three weeks later, ranking at only 12% on Google Trends compared to Irma. Yet Jose's long and drawn out journey lasted longer than any other Atlantic hurricane in 2017; in fact, Jose was the longest-lived Atlantic hurricane since Nadine in 2012. Nadine though was just a wispy bunch of mostly disorganized clouds that maxed out at just Category 1 status. Jose was much stronger, gaining Category 4 status with peak winds at 155 mph. Still, Jose, being a very impressive hurricane in 2017 that rivaled the power and duration of Harvey, Irma, and Maria, got little attention for one main reason: it never made landfall. Jose did though send epically sized surf to the U.S. east coast, which many surfers won't soon forget.

After making the usual trek from the east- to west-Atlantic, Jose brushed by the northern fringe of the Caribbean islands, causing just 0.5% of the damage left from its predecessor, Irma.[111] It was after Jose left that region though that things got interesting.

By September 15, 2017, Jose was blowing 60 mph winds while moving fairly slow at just 8 mph, located well north of any Caribbean island that could be affected by this storm. Being able to sit and spin for a while in the warm, open ocean, Jose was able to kick up seas in the 25-foot range — plenty enough energy to produce waves that would reach the U.S. east coast — as it stayed a little over 700 nautical miles to the east of Miami, and about 450 miles to the southeast of North Carolina's Outer Banks. Jose though wouldn't make a run for Florida as hurricanes often do; instead, Jose had its sights set on the Carolinas.

By September 17, 2017, after meandering in the open ocean with no clear direction, at one time staying nearly stationary, Jose was able to pump up seas reaching 40 feet when only 250 nautical miles from the Outer Banks. Aiming swell energy directly at the Carolina coast could result in breaking waves that could reach 25 feet. On September 19th, the Oregon Inlet Buoy sitting off the coast of the Outer Banks saw maximum sea heights coming in around 25 feet, but wave energy diminished once reaching the Outer Banks' shores. Jose's direct course to the Outer Banks was brief, and the peak swell short-lived. But for a couple of days, swell hit the Outer Banks with surf-worthy conditions, thanks to Jose staying far enough out to sea to send surfable waves with much less stormy chop. A few days later Jose would be nothing more than a memory for those who braved its waves, but for most everyone else, it was never even given a thought.

Hurricanes only become memorable when they hit something. Once they do, the finger of climate-change-blame gets pointed at the death and destruction left in their wakes. Harvey, Maria, and especially Irma, got all the attention during the 2017 Atlantic hurricane season as they made landfall and impacted humans lives. Yet these hurricanes, along with Jose, were only four storms out of that year's 17 to gain strengths of Category 4 or 5. This pales in comparison to the busiest Atlantic hurricane season twelve years earlier, when the 2005 season had 28 storms — 4 of them being Category 5. Katrina was one of those 2005 storms, as was Wilma, which is ranked as the most intense tropical cyclone ever recorded in the Atlantic basin. But this was, in many ways, nothing really all that new.

The 1933 Atlantic hurricane season was hyperactive with 20 storms, 5 of which were Category 4 or 5. In fact, the 1933 season produced the highest Accumulated Cyclone Energy — a sum of a season's storm strength — ever recorded. And back in the 19th century, 35 years after the Great Flood in California, the 1887 Atlantic hurricane season saw 19 storms during a hyperextended period spanning from May through December of that year. Recording intense hurricane activity for more than a century in the Atlantic is the primary reason why the IPCC and NOAA do not establish links to these storms and human-induced climate change. While one can sit and dispute that all they want, the fact is that, no matter how you slice it, there is a perception problem.

Until a hurricane makes landfall, it barely gets any attention. News crews and weather forecasters provide blow-by-blow coverage as a hurricane travels across the Atlantic, but there is very little interest by anyone until a storm hits homes. From a scientific perspective, this is unfair: we should consider the totality of each hurricane season and not just cherry-pick the storms that just so happened to make landfall. By selectively choosing to give our attention only to land-falling storms does not provide a completely picture of what's going on in the world's oceans; moreover, the *paths* that hurricanes take are not necessarily signs of changing times.

Hurricanes in the Atlantic take various paths depending on a few different elements that, like all things weather-wise, fluctuate widely. First, storms are guided off Africa just north of the equator until they reach a midway point across the Atlantic. From there, a gatekeeper of sorts decides where they'll go. Spinning clockwise, a large area of high pressure centered in the mid-Atlantic known as the Bermuda-Azores High, helps to direct hurricanes to the east as its southerly fringe touches the tropical latitudes where incoming hurricanes lie. But like all areas of weather pressure around the globe, the Bermuda-Azores High wobbles to and fro — if it nudges north a bit then it spins storms more toward the east coast of the

U.S., but if it moves a tad to the south, then storms are guided into the Gulf of Mexico (like Katrina and Harvey). The Bermuda-Azores High, while having a major deciding factor on where hurricanes will go, has little to nothing to do with the higher than normal ocean water temperatures that fuel hurricanes. But it has everything to do with where they'll end up.

In September 2017, the Bermuda-Azores High expanded wide and far with high pressure extending well into the tropics. A year earlier, in September 2016, the Bermuda-Azores High was about half that size, and was weaker. There were other differences as well between those two years with 2017 having a strong trough of low pressure over the east coast, which would put a southerly force on the storm track. So as hurricanes came along the tropical latitudes in 2017, areas of pressure to the north kept these storms along a storm track that, instead of just skirting slightly north of the Caribbean islands, dropped just a tad farther south and plowed through them.

Surf on the east coast though served once again as a bellwether of things to come. Waves breaking across North Carolina's Outer Banks and many other spots along the east coast of the U.S. in September 2017 were signs of stormy activity far out to sea. Yet the perception of what was unfolding was mostly hidden from view, and media interest. But after Irma and Maria's media storms, attention was diverted away from something that lie in wait on the west coast. Something else was being cooked up by Mother Nature, and attention would soon shift back to the west.

The Calm Before

California is a land of deception. It's not so much that it's the home to blockbuster movies and boob-jobs; it's the climate that surprises many who visit — or live in — the state. Like sirens using their beauty and song to deceptively draw sailors into rocky coasts, the green winter grass blanketing California's hillsides in the winter can be a misleading lure toward disasters that will likely arrive two or more seasons away. California's winters are a false reality, displaying lush foliage during a brief (and usually light) rainy season, that for the remainder of the year turn into desiccated fields of fire fuel.

It's great news when it rains in Southern California — reservoirs fill and homeowners turn their sprinklers off. Winter can be a great time to take advantage of California's mild climate with nature hikes, mountain bike rides, and other activities around the temporarily emerald landscape. At times, in January and February, California can resemble Bavarian mountains with thick green meadows of tall grasses and colorful wildflowers. California's grasslands come to life in the winter, as though spring had sprung in the Alps. But it's a ruse.

Once summer kicks in and the rains stop, the browning of California begins, but with little concern. But once summer ends and the windy season begins, what was lush and vibrant, then drab and golden brown, can next turn deadly. The more rain California gets in the winter, the harsher it can be in the fall — also commonly known as fire season in California.

With 2016-17 being a boon for rain in California, drought-ending excitement masked what should have been apprehensions for the coming seasons. Overlooking the dark side of California's winter rains was understandable as the abundance of precipitation that season was record-breaking, and attention had been primarily focused on drought-relief for many years prior. Rain now made the headlines. It brought a sense of satisfaction as well as hope that Governor Brown's state of "little green lawns" was in much better shape than it had been. Across the Northern Sierra Mountains, the 2016-17 rainy season brought an astonishing 94.7 inches of rain — a new record. San Joaquin Valley farmers rejoiced as that region received an overall average of 72.7 inches of rain. The state's snowpack rebounded and lakes were filled. By the end of September 2017, only 8% of California was in drought compared to over 83% of the state the prior year.[112] But we can't forget that the Golden State is not always what it seems, and too much of a good thing can inevitably turn bad.

California has an unwritten law of nature that can catch its residents and visitors unaware: What sprouts up will eventually burn down. Grasses and brush that once greened-up the hillsides later become fuel for the next fire season. With California being overly hydrated during the winter of 2016-17, the approaching fall and winter would unfortunately prove this rule true.

Fire

Blow, blow, thou winter wind
Thou art not so unkind...
Thy tooth is not so keen,
Because thou art not seen,
Although thy breath be rude.

~William Shakespeare

Every year California burns — not entirely, but partially. It's a cycle that's gone on well before people populated the region: winter rains bring growth that dies in summer and then burns later in the year. Sometimes it takes decades, but almost inevitably most areas of California will experience some kind of wildfire at some point. Same goes for a good part of the American West; in fact, some ecosystems depend on periodic fires to promote diversity while cleaning out dead plant material. Ponderosa Pine forests, for instance, tend to burn every 25 years or so around the American Northwest as the thick, protective bark of these pines has adapted to ground fires burning dead needles, thus leaving the treetops pretty much unharmed. There's also the effects of fire-stimulated flowering and seed burying that benefits from ground fires in Ponderosa Pine stands as well. The robust, thick bark provides the Ponderosas resiliency from the semi-regular ground fires, thus further allowing these tall trees to withstand fire as they benefit the surrounding environment.[113] The same goes for Chaparral, common in Southern California, which benefits from 25-year fires that sprout fresh growth of native plants. This love-hate relationship between plant and fire is a cycle so old that the environment has not just grown accustomed to it — it depends on it.

We though, have a bad habit of building our homes near lush brush and forests, believing we can control the forces of nature, or that there is no danger hidden in its beauty. It's an interesting deception that leaves us in a state of disaster denial: In California you can buy a home that backs up to open-space, but insurance companies either drop your coverage or sky-rocket your premiums, which should serve as a sign that you are asking for disaster and chancing the fate of fire. Yet costs can't seem to deter the innate, human desire to live near nature, no matter how dangerous the boundary may be between us and the tinderbox of open-space and forests we may want to live near. As a preventive measure, most California cities require land to be cleared within a 100 foot radius of your home, known as "defensible space". But when a firestorm approaches with flames many stories tall, that neutral zone is often not enough.[114] In fact, that perimeter is not necessarily a dry moat to protect your home; instead, it's more of a means to allow firefighters access to battle a blaze that would threaten your property. It's not a safety feature per se; it's just a last-ditch effort against fire's inevitability.[115] This defensible space can make a difference, but it's efficacy is too often perceived as a panacea against fire. So no matter the results, no matter how disastrous, we just keep building on the dividing line between hell and high ground.

Wildfires, overall, have become worse in recent times, rivaled only by hurricanes — two disaster-types linked to El Niño and its aftereffects. A great deal of the recent higher costs of these disasters is due to where we choose to live and build, but El Niño and its trailing effects have played a harsh role as of late.[116] Of all natural disasters in the United States since 1980, Hurricane Harvey in August 2017 was the second costliest ($125 billion), preceded only by Katrina in 2015 ($161 billion). Working your way down this list, once passing hurricanes of recent times like Maria in 2017, Sandy in 2012, Irma in 2017, and a few earlier storms including Andrew in 1992, the Western Wildfires that ravaged California in the summer and fall of 2017 come in at #15. Coming in at #10 is the drought and heatwave of 2012 — when La Niña was in full swing heating the North Pacific "Blob" that would eventually join forces with the mega El Niño of 2015-16.[117] The chain reaction of record-breaking events was now coming full circle: El Niño brought rain, followed by record-breaking rain from atmospheric river storms during the neutral Niño winter that followed, thus providing an overabundance of fuel for what would become a record-breaking fire season in California.

Following a brief intermission during the summer of 2017, Act 3 was about to get underway. Fire was about to march full force through the fall and winter of 2017. The cost would be staggering. Twenty-six years earlier the Oakland, California firestorm of 1991 caused $6 billion in damage, and the California wildfire of 2003 came in at about $5 billion. Earlier western wildfire events came in at $2-3 billion. These though were just fractions compared to 2017's western wildfires, which cost $18 billion — the Thomas Fire, by the way, was only 2% of that.[118] Yet before the American West went ablaze in 2017, evidence surfaced a couple years earlier showing things were getting worse.

A study published in the journal *Science* in 2015 by researchers from the U.S. Forest Service, UC Davis and Berkeley, University of Washington, Northern Arizona University, and The Wilderness Society, found that wildfire size, severity and frequency had been increasing globally. Fatalities from wildfires were rising, as were taxpayer-funded costs to fight these fires.[119] Their study showed that 98% of wildfires are quickly suppressed before reaching 300 acres in size; however, the other 2% burn to extremes, sparked during severe weather conditions. This seemingly small 2% of all wildfires accounts for an astonishing 97% of all fire-fighting costs and total areas burned. It's that small fraction of fires that are of the most concern, not just from the statistics of cost, but also the cause.

While climate change can be a factor, more news from the journal *Science* in 2017 showed that *we* are actually the ones to blame for the flames — not so much from our carbon emissions influencing climate change, but from our own actions and accidents.[120] As it turns out, once rummaging through a stash of government databases housing over 1.5 million records, that study in 2017 found that 84% of all wildfires over the past couple decades were started by people; the remaining 16% were from natural causes like lightning. Of the 84% of wildfires caused by humans, 25% were sparked from burning trash and debris, and a large portion were from such things as arson, campfires, and smokers. This particular study was rather timely.

In 2017, California experienced the most destructive wildfire season on record. While many previous fire seasons in the American West had been in remote areas, 2017 hit very close to home for many folks in Southern California. I for one, living in suburban Thousand Oaks — surrounded by a concrete maze of homes, streets, and buildings — witnessed at one point smoke from four active fires surrounding the area. And for a couple of weeks we watched, from our house, fire and smoke in the distant Los Padres National Forecast with flames so high that some neighbors panicked and called 911 thinking the fire was getting close — it wasn't, it was just an optical illusion from intensely tall flames 15 to 25 miles distant, and moving away from the region. Being near a heavily populated region, the record-breaking Thomas Fire placed a larger than usual population into a state of fear — and rightly so. I'll revisit the Thomas Fire later in this chapter, as it will then lead into Act 4 of our story.

No matter how you look at it, wildfires, especially in the American West, have been on the rise. Wildfires though are nothing new to the West, and as we look back over recent times, we can see history repeat itself — but with a twist.

Yesteryear

The Great Flood of 1861-62 has numerous parallels to recent times: El Niño in 1860 preceded a neutral season a year later that brought heavy rains from atmospheric rivers, which was then followed by drought. This is eerily familiar with recent times: After El Niño pounded the California coast with monstrous surf in the winter of 2015-16, it too went into à wet, neutral-Niño winter the following year, which was then followed by the rain-scarce winter of 2017-18, which dried-up the landscape. The similarities between the 1860s' events and the period from 2015-18 are uncanny, except for one thing: Fire. But something equally sinister was in store, in a proportionally biblical fashion.

Despite the copious rain and subsequent growth from the heavy rains of 1861-62, there is no historical record of any significantly large, widespread wildfire in California's history that followed. Even though California went into a notable drought following the Great Flood, the U.S. National Park Service's list of large fires is void of any entries during that time. There were two major fires in Oregon, one in 1865 and another in 1868 that burned 1.3 million acres total.[121] And there was a much smaller, ten-square-mile fire in Shingletown, California in 1870. But no major California wildfires were reported in the early- or mid-1860s following the flooding rains of the Great Flood during that neutral Niño winter of 1861-62. This may be from a lack of records, since the City of Ventura — which took a major beating from the Thomas Fire in 2017 — had just exited its Mexican period that spanned from 1821-48. Ventura was starting to flourish in the 1860s and population was rather low, about 1,800 residents during the time of the Great Flood compared to about 110,000 now.[122] Los Angeles also had a sparse population back then of just over 4,000 people, which pales in comparison to today's nearly 4,000,000 residents in the city alone. Surrounding areas bring those numbers even higher. Los Angeles also didn't have a fire company until 1869, so it's possible that fires occurred and went unnoticed; however, with major wildfires reported in Oregon but no mention of any major wildfires in California, it's quite likely that the human factor to spark the fires was missing back then.

Nevertheless, a severe drought ensued following the Great Flood of 1861-62 that desiccated the West, and contributed to the massive Oregon fires in 1865 and 1868. During this drought, many in the American West thought they were witnessing the wrath of an angry god as biblical-like plagues hit the region in rapid succession: First came the rains, then came the floods, the fires, and then, locust.

Two severe droughts got underway following the flood of 1861-61: one that revived itself by 1863 and again in 1870 as El Niño swung to its opposing extreme, La Niña. Following the neutral-Niño period that brought the rains in the early 1860s, La Niña — known for dry conditions in the American West — dried-up the region. At the University of Columbia, scientists reconstructed rainfall estimates that confirm this, based on tree ring reconstructions (taking a cross-section of a tree and measuring the thickness between each ring to tell how well it flourished during a given period).[123] The resulting models show an incredible drought that makes anything in recent times look rather moist. Compounding the problem in the mid-1860s drought was a population boom into the American Plains and West that overstressed limited natural resources. Native Americans, for instance, were migrating west as European settlers came to the eastern states. Yet some European settlers were also moving west, accompanied by an abundance of livestock that competed for natural resources as well. It was a recipe for ecological collapse.

While the American Plains and West were undergoing a natural cycle of drought, migrating humans and their livestock made it worse. These exacerbated droughts became prime conditions for a natural pest, the Rocky Mountain Locust, a now-extinct species of locust that once ranged through the American West. This small grasshopper, which traveled in swarms of many millions, could not be contained as human migration through the continental United States made a bad problem worse. As drought further deteriorated the region, the locusts thrived.

It's estimated that the Rocky Mountain Locusts caused $200 million in damages between 1873 and 1877 when they ravaged crops already stressed by drought. These hungry and highly adaptable insects ate everything in their path — not just crops, but also leather, wood, sheep's wool, and even people's clothing. Numerous, inventive attempts to eliminate these massively swarming pests did little to deter them. Farmers used gunpowder; fires were burned in contained trenches; and even sucking them up in vacuum-like contraptions did little good. Nothing worked. The locust kept coming, reminiscent of Old

Testament plagues on Egypt. Some folks even tried eating them, coming up with recipes that would pan-fry them in butter — a sad attempt at trying to make the best of a bad thing.[124]

The plains states were hit especially hard as the severe drought spread across much of the U.S, hitting the heartland with a flying, swarming, hopping, and non-stop eating, knockout blow. As swarms comprising millions of locust would descend on a farm, the sky would turn dark, putting a moving, gray-green screen between the sky and the ground, eventually pelting rooftops like hailstorms. Yards and fields quickly turned into moving masses of hungry, flying, jumping bugs. Those brave enough to venture outside cinched their pant legs with string — a necessity if you wanted to make it to your well for water, or outhouse for relief. But nothing could stop the swarms.

Once making landfall, the locust assault targeted the entire area that it blanketed. Locusts cleared fields of crops, stripped trees bare of leaves, wiped out grass, ate wool off of live sheep, devoured the leather harnesses off horses, and even ate paint off wagons, wooden handles off farm tools, and blankets and quilts that farmers placed over their crops in an attempt to protect them. Railroad tracks became slippery hazards as trains would end up squishing millions of slow-moving locusts that had parked themselves over warm tracks during the day, but then cooled at night. The attacks would last for days. In the end, the biggest rush of locusts in 1874 is estimated to have infested two million square miles of the U.S.[125]

For some reason or another, the Rocky Mountain Locusts are no more. They went extinct shortly after the devastating drought in the 1870s. It's thought that the growth of farming disrupted their cycle of swarms, eventually leading to their demise. If we had to deal with locusts in the American West and plains states today, one might think that life would be harder. Yet although the swarms of eating machines have vanished, the things they consumed have flourished, and fire that can burn them is now a different yet equally serious issue.

As mentioned earlier, we know that the greatest cause for devastating wildfires in the American West (and many other places) is from the hand of humankind. During the day-and-age of the Rocky Mountain Locusts, a less-dense, people-population meant a lower chance for accidental fires that would subsequently spread into dry-fuel infernos. But other factors are in play today that make California prime kindling, placed in the crosshairs for a wild-west firestorm.

California's fires are fairly seasonal, fueled by dry, offshore winds that tend to be strongest in the fall, and sometimes winter months. In the northern part of the state they're known as Diablos, and in SoCal they're the Santa Anas. These winds have been a part of SoCal's weather for at least 5,000 years, noted by the Native Americans who lived around what is now Los Angeles.[126] Winds in California are nothing new; they were around long before the locust swarms. But things are different today.

California, especially SoCal, was — and still is to some degree — an inhospitable place. Were it not for air conditioning, many wouldn't be willing to swelter under a scorching summer sun while watching their crops wither away from the lack of rain. While some agriculture in the state was able to utilize the limited natural waterways and resources in the 18th, 19th, and early 20th centuries, it wasn't until there came word of a proverbial promised land, bathed in heavenly weather, void of rain, that the masses flocked to the Golden State. Florida experienced something similar when developers began luring retirees from northern states to the warm climate of Florida, while not fully disclaiming that some of the world's largest hurricanes will level large portions of the state from time to time. In California, the lure of a sunny paradise resulted in a population boom. And like all things in nature, it had consequences.

More people live in California (and many other places in the American West) than ever before. More homes are located near open-space, primed for fire weather that is more often than not caused by the effects of the rising population. Yet while impetus for the ravaging wildfires of modern times is often human-induced, it's what literally stokes the fires that can be more concerning — and was especially so during the fall and winter of 2017. California can get winds that blow so brutal that they've been likened to a demon. In 2017, it sure seemed so.

Fanning the Flames

Santa Ana Winds, which through a sordid history of word-craft morphed from what earlier residents of SoCal referred to as Santana Winds, is more akin to its hellish — not hallowed — namesake.[127] In Northern California, similar winds earned the name Diablo, giving these winds an analogous, demonic moniker. No matter what you call them, these winds occur at such a regular, annual interval that a lot of Southern Californians refer to the four seasons as Rain, Summer, Wind, and Fire. The last is accentuated by its former: Wind.

The Santa Anas, though rather vicious and not usually the preferred weather in SoCal, are a meteorological marvel created from — and strengthened by — a mix of weather, geography, topography, and basic physics. During the summer, California is bathed in sunshine under a large, almost permanent dome of high pressure that spreads far and wide, even out into the Pacific, and sometimes as far north as Alaska, as far south as Southern Mexico, and as far east as the Midwest. Being the 800-pound gorilla on the weather map each summer, this massive, summertime high pressure system rules the American West. Very few low pressure systems (often associated with stormy weather) are anywhere to be found; high pressure is the king of California's summertime castle. But as fall approaches, winter weather starts to stir across the North Pacific, giving birth to low pressure systems that move toward the west coast of the United States. High pressure over the American West though stays put, which can cause a clash of weather titans.

Weather circumnavigates the globe mostly from west to east as stormy low pressure systems hitch a ride on jetstreams that flow in that direction. As fall gets underway, more frequent and eventually stronger lows forming over the Pacific take the jetstream's ride to the west coast of the U.S., and often continue through the continental United States. Like a pack of wolves surrounding its prey, low pressure systems encompass the large, western area of high pressure, creating an interaction between the two known as a pressure gradient. This difference in pressure causes wind to flow from the high to the low pressure systems, and the difference in pressure between these pressure extremes determines just how strong the wind will blow. In Southern California's fall season this can get dicey, real quick.

While the pressure gradient between these interacting high and low pressure systems will kick up wind, there are a few other things that come into play to help build the bluster. No matter how strong, weak, or otherwise unique a high-low pressure pattern can be, during California's fall season the high pressure systems tend to setup over the Great Basin — spanning the entirety of Nevada outward into the fringes of Utah, Oregon, Wyoming, Idaho, and the northern reaches of the Mojave Desert. This guides the Santa Ana winds over Southern California into an offshore pattern, blowing winds from land (the Great Basin) to sea (SoCal's coast). During the summer, this pattern is reversed as there is little influence from pressure gradients then, so as summer afternoon heat over land rises, the cooler ocean air comes ashore to fill in the void, thus creating an onshore breeze each afternoon in the summer. But in the fall, with the pressure gradients setting up between the Great Basin's high and incoming Pacific lows, winds blow in the opposite direction.

With the winds coming from inland deserts and otherwise warm locales, it's understandable to think that the heat associated with Santa Ana winds is being drawn down from the deserts. This, however, is not what warms the winds. Instead, it's compression that occurs as the Santa Ana winds blow from high elevations inland down to sea level. These *katabatic winds* (downhill flowing winds) occur elsewhere around the globe, but in California they are rather unique. As the winds descend from the mountains and higher elevations, ever-increasing pressure at lower elevations compresses the air being blown, which heats it up (known as compressional heating). So as an inland area of high pressure interacts with incoming Pacific lows, offshore winds develop that heat areas hundreds of miles away. In fact, you'll often find beaches warmer than inland areas during a Santa Ana; for instance, during a strong Santa Ana in October 2017, Point Mugu, along the Ventura County coast, had a high temperature of 88° yet Lancaster, much farther inland, was just 74°.[128] This is a far cry from summer months when areas like Lancaster can easily push triple digits while the coasts, under an onshore flow, barely break 70°. Back on June 21, 2017, when California was under yet another heatwave, Lancaster was baking at 107° yet Point Mugu was thirty-two degrees cooler, just 75°. By the fall though, when the Santa Anas start to blow, hot devil winds blow from land to sea.

Santa Ana winds can also be stronger near certain portions of the coast. For instance, during that same, October 2017 Santa Ana that had the SoCal coast much warmer than inland areas, winds were blowing at 45 mph at Point Mugu, but Lancaster was weaker at 35 mph. As those downhill winds heat-up and head to the coast, they also get funneled through various canyons and passes. One of the prime areas for this is located between the San Gabriel and San Emigdio mountains, which directs offshore winds smackdab through Ventura County — home to the infamous Thomas Fire, which I'll get to in a bit. Acting like a wind tunnel, the lower elevations between these two mountain ranges — where California State Route 14 runs — is notorious for funneling, and then further compressing Santa Ana winds as they approach Simi Valley, then Thousand Oaks, Camarillo, and eventually the Ventura County coast. Although surrounding areas feel the winds as well, this "I-14 Corridor" puts much of Ventura County in the path of exceptionally strong Santa Ana winds.

There is though another feature to the Santa Anas that is often overlooked, but can have astonishing results: temperature. Sure, the Santa Anas heat-up as they descend from high elevations to the sea, but the high pressure system besieged by the incoming Pacific lows that kicks off a Santa Ana windstorm can add a double-whammy to the mix. Sometimes, as fall gets underway, the

Great Basin high gets supplanted by an even greater area of high pressure that drops south from the Arctic, bringing extremely cold temperatures along with it. When this happens, you not only have pressure gradients whipping up the wind machine, you also have *temperature* gradients intensifying the winds as fast moving, descending, cold Arctic air butts-up against warmer air over the American Southwest. Sometimes too, a low pressure system can enter into the mix, which can also draw down frigid Arctic air. Either way it creates intense temperature gradients. When pressure- *and* temperature-gradients are in play, Santa Ana winds can be utterly violent.

Like many things in nature, complexity often creates the most severe conditions. When it comes to Santa Anas, there's an added feature that can tip the scales. Besides the pressure between the highs and lows, there's the matter of pressure throughout the various layers of the atmosphere. For Santa Anas, the concern is when dominating high pressure exists not just in the upper atmosphere, but near sea level as well. Pressure systems don't always align, but when they do, the Santa Anas become a force to be reckoned with.

So to make a perfect Santa Ana wind storm, you need strong surface *and* upper atmospheric pressure gradients, as well as strong temperature gradients. To make a perfect firestorm, you need a perfect Santa Ana and a spark, which is often linked — one way or another — to humankind (something we'll see more than once in this chapter). But can Santa Anas and the fires they influence be linked to the chain of events that were kicked off a couple years prior by the record-breaking El Niño? The short answer is, Yes. The long of it though is a bit more complicated.

The Niño Connection

If you look hard and long enough at anything, you might see patterns develop. But correlations are funny things. For instance, just because storks land on rooftops doesn't mean the town's women will have babies. Storks land on roofs and women have babies. Both instances are true, but not connected. The same goes for many things regarding weather and climate.

It's been difficult to nail down a direct link between El Niño (or La Niña) and California's Santa Ana or Diablo winds. But a study in 2012 by researchers at UCLA and NOAA found a convincing correlation between La Niña and increased Santa Ana activity.[129] This makes sense, since during a La Niña a large area of high pressure dominates the Gulf of Alaska, bending the jetstream and winter storm-track to the north of Southern California. As it does, incoming Pacific stormy lows, while being too far to the north to bring rain to SoCal, run right into the upper path of that massive area of high pressure that sets up shop over the Great Basin in the fall. So when Niño goes neutral, heavier rains tend to come from atmospheric rivers off the Pacific, which in turn accelerates the growth of brush that eventually dies and dries, which then becomes susceptible to wildfires whipped up by strong Santa Anas indicative to a La Niña that would follow that wet season as well. It's a long chain of events — from El Niño to rain to growth to La Niña to winds, and finally fire — but it is one that fits in a comfortable, natural kind of groove with no great stretch of the imagination. And studies are starting to back that up.

This then raises the question of whether Santa Anas, and the fires they stoke, would become worse during climate change, which would add yet another link to this long chain of Niño-to-wind kind of correlations. Surprisingly though, the answer is, No.

The long links extending from Niño to winds is well established, but additional research by UCSD Scripps disagrees on the idea that climate change would worsen winds (and subsequent fires) in the American West — and they're not alone. Research by UCSD posits that in a warming world the high pressure system that sets up over the Great Basin would warm, thus losing that one important element mentioned earlier: temperature gradients. Without this temperature gradient in place to strengthen fire-spreading winds, this would lead to less, not more, Santa Anas, or at least weaker ones as the world warms.[130] You might think this counters reports by the experts in the field of climate change, the

IPCC. But, while IPCC assessment reports warn of wildfires in some parts of the world, they don't mention much for the American Southwest — the home to Santa Ana winds.[131] In fact, while there is great concern that global warming could increase beetle infestations that could in turn kill trees, turning forests into dry kindling primed for fire,[132] nothing is mentioned about wind for the American Southwest.[133] Moreover though, the latest IPCC report (AR5) states "low confidence" in attributing changes in drought worldwide to human influence since it is difficult to distinguish decades-long natural variability in droughts to long-term trends.[134] This isn't a denial of climate change; it's just another example of why you can't blame everything on it. Correlations truly are fickle and funny things: climate change is real, but not everything is a manifestation of it.

That though doesn't stop people in high places from saying otherwise. Working on assumptions, outdated information, and impressive mental gymnastics, California's Governor Jerry Brown, in December 2017 said the wildfires burning that season were the "new normal", and that climate change was to blame.[135] Brown was not alone in his climate change suspicions regarding the recent fire season, but he wasn't exactly accurate either. Climate change can be attributed to many things, but not specifically to the SoCal fires of 2017.[136] But as we've seen time and again, once an idea on climate change, no matter how old or recently superseded, gets stuck in one's mind, newer science gets ignored. Dogma often falsely transcends science, and we all know that you can't teach an old dogma new tricks.

Climate aside, weather is another thing, and the link to El Niño and the fire season that followed is not just a theory; it's a wet, hot, windy fact. Once El Niño wanes, Santa Ana-fueled fires are linked to the diametric swing to La Niña. And when there's an abundance of brush to burn, a state of emergency soon follows.

The Surf Signal

Watching waves across our world's oceans can reveal what's happening in other parts of the planet. As storms thousands of miles away rage and kick up waves, that energy will eventually reach land. When it does, it says something about upcoming weather and how a season is shaping up. The frequency of swell events, where they originated, and the size of the waves can reflect the effects of El Niño, La Niña, and even neutral Niño seasons. During the shift from El Niño 2015-16 to a neutral Niño by 2017, surf in Southern California served as a beacon for the next phase of weather, which could lead to an overly-active fire season.

During summer months, California's south facing beaches (especially in SoCal) enjoy mostly two types of *swells* (the name given to wave events that usually last a day or longer). The more common kind of summertime swell is from waves generated by storms 5,000 or more miles away in the Southern Hemisphere. Being winter in that part of the world, large storms spin in the lower latitudes off Antarctica, generating seas that often reach 30 to 40 feet, and sometimes more. As these storms whip up high seas and drift north off Antarctica, swell energy is directed toward California (as well as Central America, Hawaii, etc.), which arrives 7-10 days later in SoCal. The other common type of swell during SoCal's summer is from hurricanes that spin out of the tropics. While most Northeast Pacific hurricanes travel toward Hawaii as cold water to their north (around SoCal) isn't favorable for their formation, and a permanently parked high pressure system over sunny SoCal spins clockwise to further guide these storms toward Hawaii, some hurricanes take a different course, move north, and send waves to Southern California. Like all things weather-wise on the planet, El Niño affects both types of storms and the subsequent surf that California sees from them in the summer. So watching surf patterns during the summer on how these two swell types behave is a way of eavesdropping on El Niño, getting an idea of what to expect for the coming fall, winter, and sometimes spring.

As El Niño gains strength, it does a lot of things around the Pacific, one of which is often overlooked: it strengthens the jetstream around Antarctica, the point of origin for many of Southern California's summertime swells. As this southern hemisphere jetstream strengthens, it tends to hold swell-making storms far to the south, allowing only a smidge of their swell energy to be directed north toward California. At the same time, when El Niño is warming the waters of the equatorial eastern Pacific, hurricanes become more active. So if a strong El Niño is brewing, you can observe a portent of potential winter conditions by seeing what is surfable in SoCal's summer. If there are fewer hurricane swells than Southern Hemisphere swells during the summer in Southern California, then El Niño is either going neutral, or is swinging into La Niña. But if there are more hurricane swells than "southern hemis", then El Niño is more than likely gaining ground, which can lead to a rainy winter. It's a surfer's almanac sort of thing — but it's not based on bird signs or joint pain; it's science...wonderful, surfable science.

During the spring of 2015, as El Niño was getting underway (with a modest signal of just 1°C, not even half of what it would reach in a few months) SoCal got rocked by a major swell that spun off Antarctica south of the Pitcairn Islands. On May 3, 2015, exceptionally long-period waves were teasing the buoys off the coast of SoCal, some measuring an unheard of 25 seconds apart. While most swells coming from this region of the Pacific generate periods in the 16- to 18-second range, the Pitcairn swell was generated by a much more intense storm. As swell filled in May 3rd, this strong Southern Hemisphere swell peaked in SoCal on the 4th, with many south facing breaks seeing waves running double overhead (10 to 12 feet on the face), with swell energy coming in with 20-second periods. Excitement was high that this could be the kickoff to a summer of fun in the SoCal surf-zone; however, El Niño would soon kill that idea.

The remainder of the summer of 2015 was lackluster for southern hemi surf in Southern California. There was a so-so Southern Hemisphere swell June 29th, when I also noted in my report on *Surfing* magazine that El Niño was gaining ground and that the long-period southern hemi swells would be much harder to come by for the rest of the summer. But not all hope was lost —

when El Niño closes the door to surf from the Southern Hemisphere, a window of tropical opportunity opens up.

As El Niño continued to strengthen over the summer of 2015, keeping the Antarctic jetstream far to the south, warm waters in the Pacific were starting to provide fuel for hurricanes. As you may recall from "The Cyclone Signal" section in the "Surf" chapter, the summer of 2015 was hyperactive for hurricane and typhoon genesis. Waters were warming across the entire North Pacific from El Niño and the "Blob". Southern California surfers saw the effects from this with Hurricane Delores bringing impressively sized surf July 20, 2015, and another major swell from Hurricane Linda hit SoCal September 11th. Blanca brought southerly swell in the beginning of June, and Oho brought even more swell in October.

A more bizarre swell combo though occurred near the end of August when light Southern Hemisphere swell mixed with swell from Hurricane Jimena, and northwesterly swell came to Southern California from Typhoon Atsani in the Western Pacific. This rare, summertime trio of long-period swells brought no complaints from surfers in Southern California. At the same time though, it was an exceptionally clear sign that El Niño was afoot: minor Southern Hemisphere swell was overshadowed by strong, tropical swell from a Category 4 hurricane *and* a powerful super typhoon. All three storms sent a well-flanked wave assault to SoCal, thanks in large part to El Niño.

As El Niño faded during the summer of 2016, swells in Southern California returned to their more normal pattern for that time of year. Fewer wave events were being sent from hurricane activity and more were being sent from the Southern Hemisphere. June 2016, for instance, saw three notable southern hemi swells in SoCal, one on June 1st (with waves slightly overhead), another June 15th (with waves well overhead), and then again on the 22nd (once again, slightly overhead). Your run-of-the-mill summertime surf was underway in SoCal with nary a hurricane swell to be had — a surf-signal that El Niño subsided.

In the summer of 2017, as El Niño went neutral and was leaning toward La Niña, that signal became evident in the surf zone around SoCal as well. While the state of El Niño's shift was evident in sea surface temperature readings across the equatorial Pacific's Niño zones, the inextricable connections throughout the Pacific provided further evidence to support a pattern shift in 2017 that could have devastating consequences. If La Niña was unfolding (and it was), then a windier than normal fall could spread fires that could burn the overabundance of brush grown during the previous, rainy winters. El Niño (and neutral Niño) had been on the mound pitching plenty of rain, and now it was La Niña's turn at bat.

The summer of 2017 was ho-hum for hurricane formation in the Northeast Pacific with no Category 5 storms and only two Category 4 storms. A Category 3 storm, Eugene, took an odd course northward up the coast of Baja, directing southeasterly swell at Southern California, but it was a fluke. On July 12, 2017, SoCal saw the only notable hurricane swell that summer with surf running a few feet overhead at some south facing beaches in SoCal. About two months later though, the Southern Hemisphere would remind us all that La Niña was coming back, with a vengeance.

Hurricane activity tends to peak in the late summer and fall months, but 2017's Northeastern Pacific hurricane season went quiet. Instead, the Southern Hemisphere woke up. After being fairly active during the summer of 2017, the Southern Hemisphere was about to send much more swell to SoCal to ring in La Niña's autumn return.

During the neutral-Niño summer of 2017, the jetstream near Antarctica began to weaken and waiver in true non-Niño form, breaking apart in some spots — including an area near French Polynesia where a large storm was gaining strength on September 28th. As this storm blew strong winds whipping up seas in the 40-foot range, it began to move north from a break in the jetstream, aiming swell at the California coast from more than

5500 miles away. Waves would arrive eight days later in Southern California.

October 6th saw the impact from this neutral-Niño, Southern Hemisphere swell, which brought waves running a couple of feet overhead to many of SoCal's south facing surf spots. The Wedge — a dangerous shore-break wave in Newport Beach, California, which concentrates swell energy into intensely large breaking waves — saw waves in the 15-foot range. This swell was the last to be influenced by a neutral Niño.[137] La Niña was about to arrive, and the weather it would be bring would soon spell disaster.

The memorable October 6, 2017 swell in California showed that El Niño had faded into the sunset, and its drier nemesis (La Niña) was soon to affect the American West. It's a strange kind of paradox when you think about it: As SoCal surfers in the summer of 2017 enjoyed mostly mild, sun-filled surf days with some larger, fun swells coming in from time to time, something malevolent was stirring in the waters in which they rode their boards. Many Californians fell into a similar state of false-security — El Niño was gone, the heavy rains that followed were absent as well, and everything *should* get back to normal. But it didn't.

California Under Siege

La Niña made a comeback starting in October 2017. Eastern Pacific equatorial waters that were breaking warm records two years earlier, which evened-out during the neutral Niño a year later, were now starting to cool off. As though on cue, La Niña entered the world's weather stage, with a bang.

On October 8, 2017 — days after the last neutral Niño Southern Hemisphere swell hit SoCal and La Niña got underway — the most destructive wildfire in the history of California began to burn: the Tubbs Fire.[138] Things were about to go from bad to beastly. A perfect Diablo/Santa Ana wind storm was just getting underway as all of nature's wind elements were coming together.

A low pressure system sitting off the northern coast of California had been kicking up localized wind swell in the coastal waters, but that low was now moving east, over land. A massive area of high pressure riding behind the low in the Pacific moved east as well, stretching into California and the Great Basin — where the strongest of Santa Anas begin to spin. Butting-up against the Pacific high, the intense pressure difference between it and the approaching low would eventually place strong pressure gradients over California — the opening kickoff for a classic Santa Ana. But this wasn't over yet; that often-overlooked element of cold air would make things worse. Things were just getting started.

The approaching low pressure system was housing a mass of cold air behind it, drawing frigid air south from the Arctic. This made for one heck of a temperature gradient with a difference of about 25° (Fahrenheit) between this frigid low and the massive Great Basin high in the upper atmosphere. As close as these two pressure systems were to each other, that difference was intense, to say the least. But that's not all: Pressure was aligning at all levels of the atmosphere as well. Everything — and I do mean *everything* — came together for a Santa Ana that would not soon be forgotten. It was a hellish way to kick off the Santa Ana/Diablo season in California.

While the Tubbs Fire started on October 8, 2017, it was the burgeoning windstorm that spread it quickly the next day. Winds reached 75 mph on Mt. Diablo, nearly 80 mph at the Hawkeye Raws weather station in Sonoma County, and over 60 mph in Santa Rosa.[139] As fire spread through the Fountain Grove area, hundreds of homes and structures were destroyed in no time. During the initial hours of the fire, with flames moving quickly from the windstorm, tens of thousands of people were forced to evacuate with little advanced notice. A mobile home park nearby, with 160 units, was completely obliterated. But as the day went on, and the winds refused to let up, it got worse.

During the early morning hours on October 9th, the fire reached Coffey Park, leaving complete devastation in its wake. Approximately 1,300 structures were burned and leveled, and most businesses in the area were damaged as well. Medical centers in Santa Rosa began evacuations, with some employees using their own personal vehicles to drive people to safety. By the end of the day, 31,000 people lost electricity and power as the Pacific Gas and Electric Company shut off natural gas to residents around Santa Rosa as a precaution.[140]

On October 11th, with streets walled by flames, the entire town of Calistoga was forced to evacuate. And as the fire continued to spread on the 12th, it was estimated that over 1,800 homes and 400,000 square feet of commercial space was completely destroyed. More than 34,000 acres had burned, and by the end of the month that total would rise to about 37,000 acres. Twenty-two people perished, and the cost came in at $1.2 billion.

The Tubbs Fire though was just the start of an overly aggressive fire season in California. There had been a number of fires earlier that year, including Fresno's Jayne Fire in April that burned 5,700 acres, and the Elm Fire in May burned an additional 10,000 acres. The Schaeffer Fire in Tulare County burned more than 16,000 acres, and the noteworthy Siskiyou-area fire in June 2017 burned over 65,000 acres. San Luis Obispo, on California's Central Coast, suffered from the Alamo Fire, which burned nearly 29,000 acres. And the Long Valley Fire in Lassen burned over 83,000 acres. These and other fires burned well before the Santa Ana/Diablo season got underway, when the Tubbs Fire would signal the escalation.

While the Tubbs Fire was raging through Napa and Sonoma counties, the Nuns Fire also took off in another part of Sonoma County, burning an additional 56,000 acres. Also on October 8, 2017, the Redwood Valley Complex Fire in Mendocino County burned over 36,000 acres. By October 9th numerous fires took off including the La Porte Fire in Butte County, the Cascade Fire in Yuba, the Sulphur Fire in Lake County, the Canyon-2 Fire in Orange, and the Pocket Fire in Sonoma, among others.

By the end of 2017, a total of 9,133 fires burned over 1.2 million acres.[141] Twenty of those fires were the most destructive wildland-urban interface fires in California's history.[142] These staggering numbers can stir thoughts of climate change to the forefront; however, as mentioned earlier, that needs to be done with care. While we know that a developing La Niña helped to influence the wildfire season of 2017 with exceptionally strong Santa Ana and Diablo windstorms, the link to global warming is a bit elusive. In fact, the fires themselves — not the weather that helped drive them — weren't nature's fault. It was us.

As those studies I mentioned earlier in this chapter show, most wildfires are caused by humans, not the natural environment. In the case of the Tubbs Fire, that inferno appears to have been started by electrical equipment. The Pacific Gas and Electric Company (PG&E) want to point that blame to a third party, but all parties seem to agree that winds likely damaged some sort of electrical equipment, sending sparks that kicked things off. Until the courts can sift through the mountains of legal briefs and battles, the *official* cause will be left as "undetermined" — a play-it-safe euphemism for "please don't sue me".[143]

Other fires in 2017 were human-caused as well like the Helena Fire[144] and the Canyon Fires[145]. The biggest culprit seems to be powerlines, which takes most of the blame for the California firestorms of 2017. Numerous powerlines were blown down during the powerful Santa Ana and Diablo windstorms, sparking the fires that the winds later blew.[146] Natural causes, like lightning, sparked some fires, like the Salmon August Complex Fire in 2017. Yet the majority — as the studies I mentioned earlier show — were started by human negligence.

Climate can certainly be linked to the wildfires of 2017, but climate *change*, is a bit trickier to use as the whipping boy. El Niño is a natural climate cycle, as is the inverse La Niña, which helped to influence the 2017 fire season. Rains from the 2015-16 El Niño helped to sprout fire fuel, as did the rash of atmospheric river storms in 2016-17 during the neutral-Niño. These rains may be distantly linked to climate change from overabundant atmospheric moisture, and perhaps summertime heatwaves from climate change can be blamed for turning the profusion of brush into desiccated fire fuel. But we have to remember that an almost identical pattern played out in the 1860s with just one exception: Fire. All other aspects of history repeated itself from the Great Flood era of 1862, the El Niño that preceded it, and the droughts that followed the rainy, neutral Niño floods. Fire in 2017 was the only missing variable from that 19th century event. Knowing that most modern-day fires are human-caused, and that areas like California have experienced a population explosion over the past 150 years, it's no great leap to say that history was repeating itself with the exception of our encroachment into nature — i.e. with powerlines capable of snapping in a windstorm. Thus it is us where the lion's share of the fire blame lies.

Either way, we know that the swings from El Niño to La Niña can set in motion a chain of events: a succession of surf to flood, and eventually fire. When that third element is influenced by the second, a fourth, less frequent element (mud) can show how nature not only just starts a disaster, but impels others to show their wrath. The largest wildfire in California's modern history did just that.

The Thomas Fire

The California windstorms during the fall of 2017 didn't just affect Northern California; SoCal got more than its share of bluster that season. The winds that helped spread the Tubbs Fire on October 9th also blew through mountainous regions in Southern California. Winds were especially strong around the Ventura and Los Angeles coastal mountains and hills with speeds reaching 75 mph. At lower elevations near the coast, winds approached 45 mph.[147] Another windstorm got underway October 19th that blew gusts to an unbelievable 127 mph at the top of Alpine Meadows, 8,843 feet above sea level near Lake Tahoe. Another round of wind made for an exceptionally hot Santa Ana day October 24, 2017 when Camarillo, California, near sea level on the coastal plain in Ventura County — next door to where I lived — had 47 mph winds and a maximum temperature of 104°, two degrees cooler than a couple of days prior, and actually four degrees cooler than an all-time record set two years before.[148] Combining 40-plus mph winds with temperatures exceeding a hundred degrees is like living in a blast furnace — a very dry, unbreathable and unescapable blast furnace. This windstorm though was merely SoCal's autumn inauguration. The 2017 Santa Ana season in Southern California was just getting started.

Other Santa Ana events got underway in November 2017 that blew the more typical 20 to 30 mph winds and raised temperatures to near 80° some days. These are the best-kept-secret days in SoCal, when late season tourist traffic drops off, allowing for uncrowded fun in the sun, sand and surf — if you can find a wind-sheltered area (and there are many on these *milder* wind days). But this wind hiatus didn't last for long; and all the while, things were getting dangerously drier.

A Santa Ana on November 22nd was quite strong, heating inland areas to near 100° for a couple of days. Humidity levels dropped into the single digits most days; in fact, the wild Santa Ana earlier, on October 9th, dropped the relative humidity to just 5%, and the heatwave Santa Ana on November 22nd lowered humidities to just 12%. The SoCal region was desiccated for more than a month straight. With such a prolonged span of high temperatures and low humidities caused by the La Niña-influenced Santa Anas in the fall of 2017, SoCal was quickly becoming a land of dry kindling, just waiting for a spark to set the region ablaze.

In late November into early December 2017, weather models showed hopeful signs of rain relief that might reverse that trend. Forecasts predicting weather patterns a few days away were showing a stormy low pressure system in the Pacific making its way from the Gulf of Alaska to Southern California, picking up moisture along the way. Precipitation became an enthusiastic, yet short-lived possibility. By December 3rd, weather models shifted away from the idea of rain, and were singing a much different tune.

Instead of moving over water to pick up potential rain, the low pressure system that models showed a few days earlier was now expected to ride northward over a ridge of high pressure in the Gulf of Alaska — a classic La Niña move. After being diverted north to inland Alaska, this low would spend its rounds of rain over the northern regions, but continue to hold onto its pressure energy as it swirled up and around the clockwise-rotating high in the Gulf. Like a classically blocked Pacific low in a textbook, La Niña pattern, this low, after spinning around the blocking high dove south toward California, with a trek that took it through California's now-desiccated interior. All the while, during the low's dry, inland trek through California's inland regions, the Pacific high pressure system that originally thwarted this low moved closer to the Great Basin as another, colder area of high pressure dropped south from the Arctic. A massive difference in atmospheric pressure was unfolding. It was all coming together, as though the fabled four horsemen of a windy apocalypse were riding toward the same, pestilent goal. A strong Santa Ana was developing, and this time it would be exceptionally serious, and long-lasting.

Everything aligned for a perfect, Santa Ana windstorm in SoCal: pressure gradients were strong, and a mass of cold air to the north would create strong temperature gradients as well. The gradients were quite impressive, with a temperature difference in the upper atmosphere of 35° between the cold air to the north and warmer air to its south. The dividing line between the two air masses was smackdab over the coastal range of California, placing a target on the back of SoCal. Pressure gradients are one thing, but with temperature gradients being this strong, and placed where they were, a severe bout of wind was inevitable.

On December 4, 2017 the Santa Ana was unleashed onto SoCal. Wind speeds in Camarillo, California blew over 50 mph, breaking a record set in 1957 as well as others in the 1960s.[149] At the Ventura coast in Oxnard, winds exceeded 40 mph. Farther south at the coast, Point Mugu's winds exceeded 60 mph, and Wiley Ridge in the Malibu hills saw winds reach 54 mph. And the winds just kept blowing, with no letup. All that was needed now was a simple spark to spread a fiery hell through Southern California. Analogous to the apocalyptic four-horsemen, the seven seals of judgement were about to be opened, igniting cataclysmic events, and torment. That spark came, and from an all-too familiar source.

Similar to the Tubbs and other wine-country fires earlier that year in Northern California, it appears that electrical equipment would once again ignite a vicious firestorm.[150] According to lawsuits filed against Southern California Edison, it's believed that Edison's construction at a site outside Santa Paula, California — where the Thomas Fire started — is to blame. And just like the Tubbs fire, everyone is staying hush-hush until the drawn-out legal battles come to a close (assuming they ever will).[151]

Starting in Santa Paula, hot, dry, offshore, easterly winds fanned the flames of this new fire, which many weeks later would eventually reach Santa Barbara. Before then it would plow through the coastal city of Ventura, destroying 500 homes on the first night of the fire alone. Built in the 1960s and 1970s, many homes in and around Ventura had roofs made of composition shingle, not cement tile like newer construction in the area.[152] Being only fire-retardant at best, many of these rooftops could not withstand the onslaught of red-hot embers being blown from the fiery hillsides to the communities below. This lesson would neither be learned nor addressed later during rebuilding efforts when the City of Ventura, providing construction guidelines to rebuild the fire ravaged homes, would make no mention of Class A rated roofing (even as a recommendation), which is far more fire resilient than Class B materials like those used in older shingled roofs. In fact, Ventura's reconstruction guidelines even allowed for wood shake shingles to be used, as long as they were "fire retardant".[153]

As houses near hillsides in the path of the howling Santa Ana winds became engulfed in flames, their embers — from burning shingles over plywood roofing and other flammable material thus exposed — added to the flying hot embers being blown from the blazing hills. The burning, flying debris of decimated homes added to the aerial bombardment of embers peeling off the nearby hills. Entire neighborhoods came under attack, not just from burning fragments of hillside chaparral, but also homes. The more expensive houses backing up to open-space were now assaulting homes across the street, and even blocks away as flaming roof shingles and various burning construction debris winged their way toward new victims in the path of the incessant, devil winds.[154]

More than 1060 buildings would eventually be completely destroyed by the Thomas Fire, with many more damaged. As winds blew at extremely strong speeds, water drops from firefighting helicopters and airplanes became next to impossible. Fire crews, struggling to contain the blaze, could not keep up with the fast moving fire; in fact, many residents in Ventura had no idea they should evacuate as orders weren't given (yet) but fire was rapidly approaching their backyards. Ventura residents woke to the smell of smoke and orange glow lighting the sky from fast-approaching flames. The Thomas Fire was moving so quickly that it became almost impossible to stay on top of it. Seeing news coverage that showed home after home going up in flames with what appeared to be a deficit of firefighters, some residents took matters into their own hands.

As the Thomas Fire began burning the city of Ventura, Brylle San Juan, from his home in Camarillo could see the orange glow from flames fifteen miles away. He, along with friends Matthew Serna, John Bain, Brandon Baker and Prescott McKenzie, jumped into Brylle's car and headed toward the inferno. Once there they embarked on Poli Street in Ventura where several houses were on fire and a nearby apartment complex was fully engulfed in flames. Firefighters were on-scene, but clearly overwhelmed. Grabbing garden hoses and buckets, the five young men from Camarillo were joined by neighbors who worked together for more than three hours, starting at midnight, to save some of the homes on Poli Street. Brylle admits their actions were, in his own words, "stupid", but with the magnitude of this catastrophe unlike anything seen before, many could just not sit back and let their city or neighbors burn.[155] For many, guilt and regret would be more painful than burns.

The residents assisting Brylle San Juan and his four friends were a fraction of citizens who found little time to leave their burning city. Fire outpaced evacuation orders, leaving too many people in harm's, unaware that their lives were in danger — it's something I'll talk about in more detail in the "Communications Breakdown" chapter.

By the morning of December 5, 2017 though, approximately 1,000 firefighters were on-scene battling the Thomas Fire. That number would grow to over 8,500 a couple weeks later.[156] But the fire was now spreading into the Los Padres National Forest on the north side of Ventura. Ahead of the flames laid millions of acres of thick brush on steep hillsides, inaccessible to firefighting crews and vehicles. Brush that hadn't burned for many decades, with chaparral now six feet high along with ground brush and dried grasses over two feet high, provided a nearly unlimited supply of fire fuel in the approximately 3,000 square miles of the Los Padres.[157] The Thomas Fire now had everything it needed to spread — unabated, undeterred, and unstoppable.

The city of Ventura to the south of the fire-line was still in the widening path of the fire, and flames eventually reached the downtown area, burning Grant Park above City Hall. Billows of smoke from not just Ventura but also the Los Padres region north of Sylmar were being blown out to sea from the non-stop Santa Anas, making for remarkable photos taken by NASA satellites in space.[158] When winds would temporarily calm (off-and-on) many days later, billows of greyish black smoke lifted high into the atmosphere, as though a nuclear mushroom cloud were rising from the region.

As the Thomas Fire continued to burn, its tall flames could be easily seen in nearby cities throughout Ventura County, including Thousand Oaks and Moorpark. Worried residents seeing flames at night thought the fire was almost on top of them and called 911.[159] The fire though was more than 20 miles away in many cases, but the flames were so high that it gave the illusion that the fire was much closer than it was.[160]

Offshore winds blew continuously in Southern California for another eight straight days, and then on and off again through the remainder of December 2017. It was exactly what the Thomas Fire needed to survive and continue to burn every bit of dried-up brush in its path. Additional efforts by the National Guard barely made a dent in the massively growing wildfire, and just three days into the fire, on December 7, 2017, the Thomas Fire had grown to 115,000 acres. But it wasn't done yet.

Four days later, the Thomas Fire more than doubled in size to 249,500 acres. And on December 17th the fire expanded to 270,000 acres. It wasn't until nearly a month later, on January 12, 2018, that the U.S. Forest Service would declare the fire completely contained, with a total burn area of 281,893 acres. The devastation was shocking.

During the Thomas Fire, more than 250,000 residents lost electricity as the fire damaged powerlines and other electrical equipment. Air quality became poor, not just from the offshore, Santa Ana winds blowing smoke over coastal communities, but through wide swaths of Southern California once the winds shifted to a more typical onshore pattern, blowing smoke back over large portions of Southern California. Dozens of school districts implemented closures over the long span of the fire, many concerned over the poor air quality that continued for weeks. Amtrak rail service was suspended while the fire was underway as fire reached not only the tracks paralleling the coast, but the coast itself. A boil-water advisory was put into place in Ventura after power outages caused the city water systems to fail.

When it was all said and done, firefighting efforts for the Thomas Fire totaled over $175 million, and it caused approximately $120 million in damages. Over 1,060 buildings were destroyed, and as of mid-February 2018, more than 26,000 tons of debris had been removed around Ventura from the fire — yet cleanup efforts were not expected to be completed for another two months.[161]

The annihilation of Ventura communities from the Thomas Fire was ominous and dire, leaving a dystopian looking landscape across large swaths of the city in the fire's wake. Views of the aftermath were horrifying and seemingly implausible. Houses north of Foothill Road were especially hit hard as they were surrounded by burning mountains on all but one side of housing tracts, leaving large pockets of peninsular hillside communities vulnerable in a virtual trap of wind and fire, leaving no escape from nature's fury. Aerial shots of the aftermath showed unexplainable scenes of blackened, flattened remnants of homes, surrounded in many cases with still-lush, green lawns and ground cover. Interspersed between leveled houses were occasional, yet rare, unharmed homes, with no apparent reason why they too didn't fall to the same fate as dozens that surrounded them. And in a cruel kind of irony, as wooden construction materials burned, trees surrounding most homes survived, leaving a sparse, green forest defining the corners of completely burned-out plots of land.[162]

Amazingly, only one firefighter died: Corey Iverson from far away Escondido, California, who was rightfully praised as a hero for his brave efforts.[163] One civilian died as well during the Thomas Fire, 70-year old Virginia Pesola of Santa Paula who perished during the evacuations as she tried to leave her home.[164] The costs of lives and property-loss would rise weeks later when the aftereffects of the Thomas Fire would bury people alive — when the fourth tragic element of this story unfolds in the next chapter.

The Thomas Fire though wasn't the only blaze burning in Southern California during the intense wind storms of December 2017. One day after the Thomas Fire started, the Creek Fire near Sylmar, California flared up on December 5, 2017, destroying 123 buildings and burning nearly 16,000 acres. The Rye fire in Santa Clarita kicked up the same day, burning over 6,000 acres and destroying six buildings. San Bernardino had a smaller fire (the Little Mountain Fire) that burned a couple hundred acres. And San Diego had to deal with the Lilac fire, which started December 7th and burned 157 structures and 4,100 acres.

Two of these fires — the Creek and Rye fires —resulted in the tragic deaths of at least 65 horses, many of them burned alive while trying to escape the flames that quickly swept through the rural areas where they were boarded. Despite frantic rescue efforts to free them, many horses, fearing the flames surrounding them, instinctively ran back into the burning barns where their charred remains were found the next day. Although valiant attempts were taken while those fast-moving, wind-fueled fires quickly encroached on the San Luis Rey Downs horse training facility in San Diego County, and the Rancho Padilla horse facilities near Sylmar, ranch-hands and horse owners had no choice but to flee and abandon all hope as their horses refused guidance away from the flames. Riddled with fear, most of the horses ran to the false-security of their stables, feeling they would be safe there, but instead met their deaths.[165]

A curious and somewhat ironic fire started December 6, 2017 near the Getty Center off the 405 Freeway in Los Angeles, known as the Skirball Fire. Started by an illegal cooking fire from a homeless encampment, the nearby, well-to-do neighborhood of Bel-Air lost six homes as fire from the homeless encampment spread up the hills to their community. There was an odd dichotomy and sad kind of irony as the extremely poor, cooking a meal, inadvertently burned luxury homes of the ultra-wealthy who were likely unaware of the proximity and plight of others nearby.

For a time, many Southern California residents lived in a perpetual state of anxiety as it seemed as though they were constantly surrounded by persistent flames and endless smoke. Each day appeared to bring news of yet another fire, and the frustrating inability to contain it. It was a wind- and firestorm season that just wouldn't quit. Santa Ana wind events tend to last for a couple of days, but the Santa Ana winds blowing the plethora of fires around Southern California in the fall of 2017 lasted for nearly a month.[166] It seemed that nothing could stop the continuous spread of wildfires as fire crews could not get any break in the action, and nothing seemed to help.

Out of all the fires that burned in Southern California in the fall of 2017, the Thomas Fire was the most salient, and resilient. Day after day the containment percentage remained in single digits as the fire continued to spread. Burning areas unheard of in the history of Ventura County — downtown regions, and even up to the ocean in some areas — people could not believe what they were seeing. When the unimaginable happens, it's hard to imagine what will happen next. Fear is the natural byproduct the follows, and in the fall of 2017 it never went away.

Fourteen years earlier San Diego suffered the Cedar Fire, which up until Thomas, held the record for most acreage burned at 273,246. The Thomas fire burned 281,893 acres. But even after the Thomas Fire was out, the problems weren't over. After the mega El Niño led to record-breaking atmospheric rivers during a rare, extended neutral Niño phase, which brought excessive growth around the hills of Southern California, La Niña met-up with human negligence, kicking up intense Santa Ana winds that fanned flames sparked by human error. While nature played a role in setting the stage, it was the hand of humankind that drew back the curtain for the tragedy to play out. But Act 3 of this story was merely building up to a grand finale. A few weeks later, in Montecito, California a horrific tragedy would unfold. Before then, as Thomas and other fires scorched California, other parts of the planet were suffering as well.

When the World Burned

Although human-caused fires were to blame for much of what burned in California in 2017, and while it may be difficult to directly link climate change to those events, 2017 also saw the largest wildfire ever to burn in British Columbia, totaling over 3 million acres of forest. It was an event taken right off the pages of the IPCC reports and their links to global warming.

Heatwaves and minimal rainfall leading up to the summer of 2017 left many of the forested areas around the westernmost regions of Canada extremely dry. Coupled with warmer and longer than normal summers, mountain pine beetles were provided longer life- and breeding-cycles that in turn killed large swaths of the northern forests — which in turn also provided immense areas of fire fuel. This was classic climate change, noted by the IPCC in their assessment reports that show this sort of cycle has continuously worsened since 1994.[167] Severe droughts with unusual extent and severity have affected not only Southern California, but also large portions of North America including Canada. Tree mortality has taken a hit, exacerbated further by higher than normal summertime temperatures, which has severely affected northern trees like Trembling Aspen, Pinyon Pine, and Lodgepole Pine. Insect infestations during these droughts — in some ways similar to the locusts in the drought of the 1870s — have caused an increased rate of forest dieback, which in turn provides copious quantities of dry, fire fuel.

In July 2017, typical of summer in North America, a large area of high pressure was parked over the American West, spreading north to Canada. A strong, stormy low was sitting just off the coast of British Columbia though, and it was headed to that region. This however would be yet another dry storm, with rain hard to come by. But this storm's dynamics were enough to spark dry lightning, which is the likely cause for the B.C. fire. Prior to this particular fire though, out of the 572 earlier fires in British Columbia during the first half of 2017, about half (258) were caused by people.[168] Most were quickly extinguished. But by the end of August, 19 different wildfires were raging and had merged into one, creating a blaze covering 560,000 acres. A month later that total would grow to 3,000,000 acres.

Elsewhere around the globe fires raged as well in 2017 with Italy and Romania burning an area the size of Rhode Island. In June 2017, 60 people died in one weekend alone from wildfires in Portugal — an additional 30 people were killed when fires reached roads on evacuation routes. Europeans called that season "Lucifer" due to the extremely warm temperatures, lower than normal rainfall, and subsequently widespread wildfires.[169]

Wildfires in 2017 also spread across large parts of Siberia, Brazil, South Africa, Chile, and New Zealand. All affected by dry, drought conditions. But most, like the Thomas, Tubbs, and many British Columbia fires, were not started by nature; they were started by people, accidentally and in some cases intentionally.

In many forested areas, burned trees give way to growth. Known as an "early seral" ecosystem, forests have a way of sprouting new plants to cover the earth after a fire. Eventually seeds from destroyed or semi-damaged trees make a comeback, growing a canopy that eventually blots out sunlight to the forest floor, killing off some of the underbrush while allowing the trees to thrive, dominantly. This though is a bit different in Southern California.

In SoCal, shrubs that occupy much of the hillsides cycle through a similar fire regime that other early seral environments endure. Although not having tall tree forests like in the Pacific Northwest and Canada, the chaparral around SoCal is accustomed to the seemingly endless cycle of growth-burn-regrowth. A durable cycle can withstand intervals from burn to burn spanning one to two decades. In fact, many of the mountain ranges around Southern California have brush that hasn't burned for many decades. When it does burn though, eventually the shrubs and limited trees return, as though nothing ever happened. This can provide a deceivingly false sense of security.

As human expansion continues to push into — or up against the boundaries of — California's natural habitats, we do so at high risk. Kinetic energy, stored in wild brush, cycles through decades-long wait periods before flaring up into blazing infernos. Like earthquakes, the question of wildfires threatening the edge of SoCal's cities and suburban sprawl isn't "if"; it's "when". Fires will rage in SoCal; it could just be a long wait until an area gets ignited. Once it does, a new threat appears from denuded hillsides. It's a threat that sits in silence until the windy Santa Ana season comes to a close, and the fires are out. Positioned in the path of rain runoff, homes in Southern California often sit in the proverbial eye of the storm. Clearing a 100-foot perimeter between human habitat and desiccated land is helpful for fire containment. But once the fires on the hillsides that meet that boundary are extinguished, rain that was originally longed for becomes residents' worst enemy. Once the flames are gone and the fire crews leave the burned hillsides around California, rain will come. When it does, something unstoppable, and potentially worse, waits in the wings.

Mud

We cannot stop natural disasters,
but we can arm ourselves with knowledge.

~Petra Nemcova

It would be comforting — though naïve — to think we live on stable, solid ground. In reality, we reside on a mere veneer of dust on rock. Most of the material that comprises our planet lies well below the surface, with soil being a miniscule fraction of the 20,000,000-foot thickness from where we plant our feet to earth's internal heat.[170] With even the thickest soils measured in hundreds of feet, the layer of dirt that we call home covers what equates to a light dusting of sugar on a chocolate truffle. When something disturbs that thin layer of deceptive dirt, rivers of debris ensue.

Out of the dozen impacts noted by the IPCC to be of the most concern under climate change, flood and mud rank together in the number one spot.[171] Hurricanes come in a close second, and water quality runs at number three. Yet the most popular and often talked about areas of climate concern — sea-level rise and heatwaves — come in lower at numbers four and five. It's a rather odd kind of paradox where the most important aspects of climate change are rarely talked about by global warming's most ardent advocates — like Al Gore. Using Google Trends once again — like I showed earlier when looking at hurricanes and headlines in the "Flood" chapter — we can see that the most dangerous and potentially costliest climate disasters get the lowest public attention. During 2017, there was more interest in sea-level change from climate change than anything else. Flooding came in second, and heatwaves third. But as far as landslides or mudslides go, interest was so imperceptible that it couldn't register anything on

Google Trends.[172] It's odd that sea-level rise — something slow-changing that spans decades if not centuries — is given more attention than flood and mud, which are of a more immediate concern. The IPCC gives flood and mud precedence, but those who preach — and profit from — climate gospel don't seem to mention it very much. Here though, I will.

Act 4 in our story of El Niño and its trailing years of extremes, Mud, ends the vicious cycle of wave-rain-burn-slide with some of the most nightmarish and devastating effects. The IPCC seems justified in placing this kind of impact at the top of their list, and anyone who has lived through a mud- or landslide will attest to that as well. Surf we can see, and know when it's coming. Fire is something that also has some level of fair warning, and can take days, weeks, or months to control — but it *can* be controlled. Mudslides though happen in an instant, with little warning and nearly no time to get out of the way. While flooding rain is seen as the culprit leading up to a slide, mudslides, while not getting as much attention, are complicated, and sometimes misunderstood.

California — SoCal in particular — has a curious geological structure that makes it susceptible for slides in many ways. Over a billion years a lot has happened to shape the high mountains and steep slopes that give many areas of SoCal its grandeur. Sitting next to sea level are rugged ridges and canyons that have been forced upward over seemingly endless eons from the same tectonic activity that sets off earthquakes around the region. Some of that rise is dramatic, taking it up sometimes thousands of feet above sea level over the past tens of millions of years, evident in areas like Fossil Trail in the Point Mugu State Park, where I often like to hike. There you will find, many hundreds of feet above sea level, imprints of sea scallops, sand dollars, and other fossilized marine life, left there a few million years ago. Wildwood Park in Thousand Oaks — another one of my favorite hiking haunts — has fossilized oysters in its caves, and I've come across ancient rock-burrowing, ocean mollusks there as well. All of these remnants of ancient sea life are located a good thousand or more feet above sea level.

As tectonic plates collided and pushed the earth's rock and dirt upwards from the sea to today's mountaintops, gravity is constantly trying to pull it back down. As dirt builds up over the mostly rocky, tall hills and mountains around California, it is bound together by plants that take root in its thin layer of soil. Building a web of flora roots helps to keep mountain soil from moving. If the plants disappear though, then disaster awaits. But in California, the flora lining the hillsides have an adaptation that makes things more problematic.

California's wild plants have evolved to survive within the rocky, thinly soiled hillsides around California, while building a resistance to moisture evaporation. These plants need to hold on — for dear life — to all the water they possibly can during the dominantly dry seasons. These desert-like plants do so with a waxy coating on their stems and leaves, which helps seal in what little moisture they absorb when infrequent rains eventually fall. While the wax-coated chaparral and shrub-lands that line the hills and mountains of California have adapted this water resistant mechanism for their survival, remnants of them, left over from a fire, can make the chance of a mudslide even higher.

When the waxy plants on California's hillsides burn, the oil-based consistency of those plants provides plenty of fire fuel that becomes darn near impossible to extinguish. But an even bigger problem is that not all of this waxy oil burns; some of it drops to the ground. This then creates an oily film over the soil, creating what is known as a hydrophobic, or water-repellent soil.[173] This, in turn, creates a virtual water slide for rain to move more quickly down hillsides, picking up rocks, trees, brush, and also dirt from areas not adversely affected by this now wax-coated soil. Something similar happened in the Christmas Flood of 1964, which you may recall earlier devastated areas of the Pacific Northwest. Severe mudslides occurred in December 1964 from ground that was also hydrophobic. But in the case of the 1964 flood, it wasn't oil drippings from burning waxy plants, but instead an exceptionally cold winter had left the ground frozen on many mountains and hillsides in Washington and Oregon, preventing rain from soaking in, and instead running off quickly.[174] Both are cases of hydrophobic soil, where rain rolls off the ground, often in torrents. In Southern California, where frozen ground is not a concern, the waxy plants are the cause for exacerbating rain run-off and subsequent flows of mud and debris.

As a general rule, once a rainy season in Southern California reaches ten inches of rain, it becomes especially susceptible to mudslides.[175] Once that level of saturation occurs, any short-lived, intense rainfall can be enough to trigger a slide — oftentimes two inches over six hours. These time scales are lengthened during dry times between rain storms, but fire changes the formula completely.

Once you burn away the plants and their intricate root structures holding a hillside's surface dirt together, and as plants' subsequent waxy oils begin coating the ground, all it takes is a quick dowsing to level the playing field — literally. Once hillsides in SoCal have been ravaged by fire, all it takes is a brief downburst to trigger extremely catastrophic consequences.

Montecito

Located just east of the city of Santa Barbara, California sits the wealthiest community in the county, Montecito, which received more than just a one-two punch from the spat of events spanning 2015-18 — it suffered a fatal blow. After flooding rains led to the eventual growth of fire-fuel, and hillsides burned in-kind, returning rain would entomb homes in an avalanche of mud and boulders — people would be buried alive. Sadly, officials knew this would happen, but mistakes and misunderstandings — by not just officials but residents as well — made a serious catastrophe much worse.

On January 7, 2018 an extensively long trough of low pressure was sagging far to the south near California, from the Gulf of Alaska to near Mexico. A low pressure system positioned itself at the very southerly tip of this trough, centered just off Point Conception, California in Santa Barbara County. As this low pressure system sat off the coast and swirled counterclockwise, it tapped into a rather robust plume of moisture to its south, packed with precipitable water. To make any storm a rainmaker, a moisture plume just needs to be in reach of a strong spinning low. In this case, the low was strong, and the moisture plume extensive; an atmospheric river pattern had setup over Southern California stretching from Santa Barbara southwest to the tropics. Rain, and plenty of it, was headed to the California coast.

By Monday the 8th, trace amounts of rain had fallen around Santa Barbara, but the storm sitting off Point Conception hadn't yet come ashore. Forecasters at the National Weather Service office in Oxnard warned that severe rain was headed to the Santa Barbara region, which could threaten the vulnerable, fire-denuded hillsides around Montecito. The Thomas Fire was no longer a concern, but the barren, burned hillsides ravaged weeks earlier by the fire now left great portions of Santa Barbara county's mountains unprotected.

At 11:30 AM the County Executive Office of Santa Barbara took the National Weather Service's warnings under advisement and issued an official press release, proclaiming a local emergency due to a severe winter storm approaching the area. That press release, in part stated,[176]

> *The National Weather Service issued a Flash Flood Watch for the county beginning today through Tuesday, January 9, with the storm intensity increasing overnight to include winds of 40 to 70 miles per hour and rain amounts of anywhere from 1/2 inch to one inch or more per hour. The risk of debris/mud flows and flash flooding is greatly increased in the areas burned in the Thomas, Whittier, Alamo, Sherpa and Rey fires. The intensity of the Thomas Fire left our mountains with little or no vegetation to prevent the slopes from sliding...the threat of flash floods and debris flows is now 10 times greater than before the fire — they can happen with little or no warning. A*

> *mandatory evacuation order and evacuation warning is in place. It is imperative that the public understands the seriousness of the situation and is prepared. Do not delay in taking action to protect oneself, family and property.*

This press release was sent to the media and the Santa Barbara County Sheriff's office, which kicked-off confusion and later led to the loss of life. Evacuation maps concerning imminent mudslides were drawn up, but a map on the county's website showed a larger "voluntary" evacuation zone than the map on the sheriff's website did. Later that would prove to be a fatal mistake, as the county map was spot on, but the sheriff's map left people in harm's way.[177] No alerts had yet been pushed to mobile devices — the press release and conflicting evacuation maps comprised the entirety of the information available to Montecito's residents. Depending on which map one viewed would determine their chances of survival. Not fully realizing which areas were under voluntary evacuation, some Montecito residents felt a false sense of security, since Thomas Fire "mandatory" evacuations earlier made the now-voluntary evacuation areas look less concerning. But there was no comparison — both hazards were equally threatening. All the while, the storm was fast approaching the Montecito region.

By 3:00 AM Tuesday January 9, 2018, the full brunt of the powerful Pacific storm came ashore. As the low nudged closer toward Santa Barbara, the atmospheric river moisture plume it tapped into got the added benefit from something known as orographic enhancement: As the stormy low circulated counterclockwise, drawing moisture into its weather front, the coastal mountains in the path of the weather front, around Montecito, stood in the storm's way. This caused the incoming atmospheric moisture to pile up and eventually lift, compressing it into a massive ball of dense moisture. Once overly laden with water, to the point of excessive saturation when it could take no more, that mass of moisture would burst, unleashing a watery hell.

A half hour earlier, the National Weather Service office in Oxnard, California pushed a cellphone alert to residents around Montecito to evacuate, immediately. But something went wrong — terribly wrong. For some reason, the alert didn't seem to go; only a handful of cellphones lit up, sounded, or buzzed. It was at that moment that the director of the emergency management office in Santa Barbara, Robert Lewin, became very concerned. Keeping a cool enough head, Lewin followed-up with their own flash-flood alert within 15 minutes of the failed emergency alert that was supposed to take advantage of a much wider cellphone audience — something I'll talk about more in the "Communication Breakdown" chapter, when I talk about as this and other similar communication debacles.

Unfortunately, the limited, second alert issued by the emergency management office was only pushed to people who had proactively signed up for official notices, or who were on social media at the time. With most residents of Montecito sleeping at 2:30 AM, this limited alert was like the proverbial tree falling in a woods — no one (or very few) heard it.

Meanwhile, the rain didn't just *continue* around Montecito; a deluge of biblical proportions let loose. Over a half inch of rain pelted the Montecito hillsides over a brief, five minute period at 3:30 AM, adding to the inches of rain that had already fallen.[178] The hillsides around Montecito, denuded of foliage and covered in oily, waxy resin from the burnt brush, were about to release an unfathomable amount of their surface. Gravity would take care of the rest.

By 3:50 AM, Santa Barbara County officials finally sent a successful alert through the federal wireless system that failed to send the National Weather Service's alert earlier. When questioned why, Jeff Gater, the county's emergency manager, said that officials originally decided not to use the cellphone alert system out of concern that it might not be taken seriously, saying "If you cry wolf, people stop listening." It was yet another fatal mistake of human error made during the most critical point of the unfolding, and quickly escalating, deadly disaster. The person who was assigned to sound the alarm, who was paid to do so by taxpayers in the region, failed to do his job.[179]

By the time the alert finally went out, it was too late; the deadly debris flow started minutes before. The hillsides above Montecito had already released unimaginable torrents of mud, along with anything that stood in its path — and more was on the way. People were being buried alive. Houses were crushed and turned to rubble. Cars were being washed down streets. Moving at speeds of up to 20 mph, mud as thick as 15 feet plowed through everything that lay before it. Within minutes, power lines were downed, causing the loss of electricity to over 20,000 people. And shortly thereafter, a 30-mile stretch of the 101 Freeway was shut down as mud two feet thick covered the roadway.

Some residents, who had grown weary of yet another evacuation — first from the Thomas Fire, and now from the potential mudslides — ignored the warnings and stayed in their homes. This too, was a fatal mistake. Despite the efforts of more than 1,250 firefighters, along with the California Conservation Corps, and the California National Guard, by January 21 it was confirmed, after finally being able to dig through all of the mud, that 21 people had died and over 160 were injured. Two remained missing.

It was clearly evident days earlier that there was potential for rain heavy enough to cause downpours. From my report on *Surfer* magazine on Sunday, January 7, 2018, I wrote:

> *...A low to our southwest will initially stir in precip to SoCal Monday as it taps into moisture from a hefty plume sitting just offshore, extending into a moderately sized atmospheric river. This is a precursor to Tuesday. Monday should see light to moderate rain for the most part. But then by Tuesday, a stormy low driving southeast from the Gulf should tap into that moisture plume, wringing copious rain out of it. Latest model runs show rain reaching SB shortly after midnight, and bringing off and on showers through the*

day Monday to much of SoCal. Heavy rain is then expected to hit LA north pre dawn Tuesday, rain heavy during the day Tuesday, and then taper off by midnight or pre dawn Wednesday. Rain totals based on today's models put SB at 2-3" of rain; VC up to 2"; LA 1.5-2"; OC 1.0-1.5"; and SD 0.25-1.25".

The National Weather Service issued other warnings as well. In fact, the day before the storm, about 60 Santa Barbara County sheriff's deputies, armed with information from the weather service, joined forces with search and rescue workers to canvas the charred communities below the potential mudflows, persuading people to leave. County officials had also issued mandatory evacuation orders for about 7,000 people near the Thomas Fire burn area, and voluntary orders were issued for another 23,000 people in harm's way. Many people though decided to stay and ride it out. But when the emergency, "get out now" alerts finally made it to the many who remained in the Montecito area, it was just too late. Had the alerts gone out earlier — as they should have —lives may have been saved.

One of the victims from the Montecito slide was Josephine "Josie" Gower, age 69. As her house became knee-deep with mud, and boulders crushed then carried away her garage, she and her boyfriend Norm walked downstairs, and together they opened the front door to look outside. As they did, a wall of mud and boulders, some as large as pickup trucks, crashed through Gower's home, sweeping Josie and Norm out the front of the house and into the melee of mud that was flowing through their neighborhood. Witnesses say that Norm, clinging to a tree, reached out his hand for Josie, and yelled her name — for hours — but to no avail. Josie Gower was gone.[180] A similar story played out twenty more times.

Along with Josie Gower, the list of those killed include people ranging from 3 to 89 years old. The official cause of death for most of the victims of the Montecito mudslide of 2018 was "Multiple traumatic injuries due to flash flood with mudslides due to recent wildfire." To pay homage to these victims, I've included their names below, in hopes that we can honor their memory:[181]

Faviola Benitez Calderon, 28 years old

Jonathan Benitez, 10 years old

Kailly Benitez, 3 years old

Joseph Francis Bleckel, 87 years old

Martin Cabrera-Munoz, 48 years old

David Cantin, 49 years old

Morgan Corey, 25 years old

Sawyer Corey, 12 years old

Peter Fleurat, 73 years old

Josephine Gower, 69 years old

John McManigal, 61 years old

Alice Mitchell, 78 years old

James Mitchell, 89 years old

Caroline Montgomery, 22 years old

Mark Montgomery, 54 years old

Marilyn Ramos, 27 years old

Rebecca Riskin, 61 years old

Roy Rohter, 84 years old

Pinit Sutthithepa, 30 years old

Peerawat Sutthithepa, 6 years old

Richard Loring Taylor, 79 years old

Two people remain missing.[182]

The cleanup efforts were arduous, to say the least, with people still digging out the following month facing seemingly insurmountable fields of mud and debris filling their neighborhoods. Water lines running to Montecito were damaged for weeks as well, including some that had been installed in the 1920s to bring water from Jameson Lake in the Santa Ynez Mountains to Montecito below. With mains damaged, Montecito residents were left with little in the way of safe drinking water. All told, there was significant damage to the primary water transmission line; there were 15 breaks to the main water line; 25 hydrants were sheared off in Montecito; and 290 water-service lines were damaged or leaking on properties as well.[183]

After the Montecito slide, the 101 Freeway was initially closed from Ventura north through Santa Barbara, and then later in a smaller section that isolated Santa Barbara from everyone to the south. During the two-week effort to clear the mud and debris choking off the main artery running through Santa Barbara, those traveling south, even to neighboring Ventura, had to either take hours-long routes inland, or pay for an ocean ferry service to go around the Montecito slide. After hauling off over 105,000 cubic yards of mud at a clean-up cost of $12 million, the 101 Freeway was reopened — two weeks after the slide. But where the mud would eventually go caused additional concern.

There were limited choices for disposing the massive amounts of mud that flowed through Montecito. You can't merely take it back from where it came, and the vast quantity of mud made for a logistical nightmare. To alleviate the cleanup efforts as quickly as possible, much of the mud and debris were taken to local Santa Barbara beaches and dumped in the surf zone. The Santa Barbara County Flood Control District was able to obtain emergency permits from the U.S. Army Corps of Engineers, the California Regional Water Quality Control Board, the California Department of Fish and Wildlife, and the California Coastal Commission to place this "sediment" directly on local beaches, untreated.[184] This, in turn, created another crisis, contaminating the beaches and ocean waters around much of Santa Barbara and Ventura counties. The Santa Barbara County Public Health

Department and the Ventura County Resource Management Agency, during their routine testing, found excessively high levels of bacteria at local beaches for many weeks. Numerous beaches were either posted and/or closed nearly a month later as convoys of trucks continued to dump mud onto Santa Barbara's shores.[185]

Besides the high bacteria levels contaminating Santa Barbara and Ventura County beaches, there was also concern that there could be other, untested contaminants including gasoline and oil from smashed cars, as well as household and garden chemicals swept away by the rivers of mud. How this will play out in the future is uncertain. While not necessarily as severe as the 2010 Deepwater Horizon oil spill in the Gulf of Mexico, contaminants and potential carcinogens from oil, gasoline, and various chemicals could have long term implications on fish, animal, and plant life. As a precaution, the Santa Barbara County Public Health Department issued a press release[186] January 17, 2018 in the form of a public health advisory, stating:

> *The storm and mudslide event caused extensive damage. As a result, unknown amounts of potentially hazardous chemicals and untreated sewage were swept into the mudslide debris that flowed through impacted areas. As people return to these areas and begin the difficult task of cleanup and recovery, they are advised to take certain measures to protect their health. They are also advised to be alert to certain health conditions associated with natural disasters, disaster cleanup, and repopulation of impacted areas.*

The advisory included health risks such as rashes, trench foot, wound infections, and gastrointestinal illness. Vaccinations were recommended for tetanus as well as hepatitis A and B. Boil-water notices were included in the advisory as well.

The amount of mud and debris was so immense from the Montecito mudslide that a month after the slide, as clean-up efforts became increasingly difficult, the County of Santa Barbara urged homeowners to try and integrate the hillside debris into their landscapes. The idea turned from cleanup, to just leave it. Director Rob Lewin — as you may recall had problems with the late-issued alerts attributing to fatal mistakes before and during the slide — said,

> *We are encouraging people to find innovative ways to keep the material on your property both in the short term and the long term, to work with your architect to identify how to improve your property with the material on site.*[187]

My first thought was, “Seriously?” It’s hard for me to imagine leaving car-sized boulders or mountains of mud in your yard to add curb-appeal. Or raising ground level around a house that could lead to rain drainage toward a home. Director Lewin, by the way, also took ownership of the evacuation map discrepancies mentioned earlier, admitting that prior to the storm’s arrival, he approved a press release and Facebook post that had discrepancies with the evacuation zones.[188] I’ll discuss additionally odd methods Lewin came up with in the aftermath of Montecito in the next chapter.

While Montecito struggled to cleanup and slowly get back to some kind of normality, residents knew that winter had only begun. More storms could come. The Pacific never sleeps.

The Near Miss

Montecito dodged a bullet a couple of weeks after the devastating mudslide in January 2018. If it had been an El Niño season, instead of one swinging toward La Niña, things could have been much worse when, on January 19, 2018, a massive Pacific storm gave Central and Southern California a close shave, and narrow escape from further destruction.

During the week of January 14, 2018, the jetstream across the North Pacific was sagging south into El Niño-like latitudes, driving the Pacific storm track out of the Western Pacific toward warm waters near Hawaii. This usually isn't the case during a La Niña winter, as high pressure would normally bend the storm-guiding jetstream far to the north, stopping storms coming out of the Western Pacific from gaining momentum. But on the week of January 14, 2018, this wasn't the case. Instead of being bent to the north to take Pacific storms to a final resting place in the Bering Sea or Alaska, the jetstream was poised to pump-up any storm that took a ride along its storm track as it guided low pressure systems southeastward to warm waters north of Hawaii. This pattern would eventually bounce back to a more, La Niña-like state a few weeks later, but for now, the middle of January 2018 was looking a lot like winters of El Niño past. This pattern wouldn't last for long, but while it did, concern grew that Montecito and other burn areas could be in for a beating.

Once the jetstream was in a storm- and surf-worthy, low latitude position, one particular storm hitched a ride on the jetstream toward Hawaii, gaining intense strength once nearing the Hawaiian longitudes. This storm continued to increase in strength as it progressed toward California since nothing stood in its way. When reaching a 2- to 3-day window for swell to arrive on the west coast of the U.S., this storm raged with seas running 40 feet, and had no signs of backing down. But luckily for Montecito and the other burn areas around SoCal, something would temporarily stop this wild, Pacific storm from battering the region.

As this monster storm continued to move east across the Pacific, a dome of high pressure formed off the California coast, which bent the jetstream to the north, taking this storm along with it. It was like a bubble formed over Southern California, protecting the region from approaching storms. Strong surf though was consistently sent to Central and Southern California from when this system was more than 2,000 miles away. Swell continued to be fired at California up until this storm got within a too-close-for-comfort, 1,100 nautical miles away. That's when it encountered the ridge of high pressure sitting off the coast of California, which thwarted the storm northward to eventually make landfall around Oregon; thus sparing the charred, mudslide-susceptible hillsides of SoCal from further erosion.

Instead of rain, Santa Barbara had clear conditions January 19, 2018. Surf was pounding the SoCal coastline, but rain was staying away from the hills in Santa Barbara County — Montecito was safe (relatively speaking). At 4:00 PM the day earlier, 20-second periods started to hit the Point Conception buoys off the coast of Santa Barbara, and by 7:00 that night waves were rocking the Harvest Platform buoy off the coast with 16-foot seas. While dangerous surf ensued the next day — as SoCal stayed dry — Oregon got a direct hit.

Even though Point Conception saw impressive, 16-foot seas (and larger *breaking* waves) from the thwarted Pacific storm, buoys off the Washington and Oregon coasts were being hammered with seas in the 40- to 45-foot range on January 18th. Breaking wave size would be bigger. Curious wave watchers flocked to the coast, and as titan-sized waves pounded the coastline of Oregon, one man, Alfredo Rodriguez Alvarez, got too close to the shore at Depoe Bay and was swept out to sea. He was presumed dead later that day, but never found.[189]

The Oregon coast received 2 inches of rain on the day the storm and swell hit Depoe Bay; another 1.7 inches January 19; and another 1.1 inches on the 20th. Had this storm hit Santa Barbara and not have been thwarted to the north on the last leg of its sojourn across the mid latitudes of the North Pacific, Montecito would very likely have been in a much more dire situation. The 101 Freeway, which had been covered in a flood of mud from the Montecito slide was still closed. Cleanup efforts had barely started around the buried communities of Montecito. And some victims of the mudslide had still not been found. If another deluge, or even moderate rain event had hit Montecito January 19, 2018, the ending to this tragic story would have certainly been much worse. The story though, may not end for some time to come.

Luckily for Montecito, February 2018 was a relatively dry month. Only trace amounts of rain were recorded. If though the normal 4 inches had fallen, the Montecito crisis would likely have continued that month. But thanks to La Niña, the storm track that month took storms coming out of the Western Pacific to high latitudes in the Gulf of Alaska, leaving Southern California dry. In a classic La Niña pattern, high pressure had setup shop once again in the Gulf of Alaska, and in its clockwise spin drove mostly dry low pressure systems south along the west coast of the United States. That pattern though shifted slightly by the end of the month, which raised a warning flag for Santa Barbara.

On March 1, 2018, the rain-blocking high pressure system in the Gulf of Alaska had moved just slightly to the west, allowing a storm being driven south on its eastside spin to extend slightly over water, picking up precipitation along the way. Weather models showed a blob of moisture hitting Santa Barbara County by Friday the 2nd, but luckily for the residents of Montecito, rain fell at a lighter and less rapid rate than expected. Yet enough rain fell to cause issues in the area: Highway 33 near Ojai was closed from a light to moderate debris flow, and portions of Highway 192 were closed as well due to flooding and debris flows there.[190] This storm was minor, but enough to create these issues from what little rain fell.

It will take years for Montecito's residents to fully recover. And it could take longer for the hillsides surrounding the city to regrow protective, soil-binding brush. Until that happens, the thoughts of rain will no doubt be cause for anxiety. Because as we and the residents of Montecito know, the Pacific never sleeps.

Big Sur

Along the central coast of California, just north of Hearst Castle, sits one of the most scenic yet perilous drives in the world; a section of California's Pacific Coast Highway that leads to Big Sur. A winding road with spectacular, scenic views, the Big Sur Highway is merely a lightly carved edge on steep mountain slopes that line a section of the California coast. Rock- and mudslides are inevitable in this region, and in May 2017, a major slide occurred that changed the landscape forever.

Just like the rest of California, Big Sur got more than its fair share of rain during the neutral-Niño winter of 2016-17 when atmospheric rivers brought an overabundance of rain to the region. Once the rainy season got underway in October 2016, Big Sur received over 6 inches of rain, and more than 11 additional inches fell before the end of the year. Things then kicked into high gear as the January to February 2017 rash of atmospheric rivers brought over 25 inches of rain in January, and over 22 inches in February. While March 2017 gave Bug Sur a chance to dry out a bit, another 5 inches of rain came in April, further saturating the soil, setting things up for a slide.

Being so close to the coast, the mountains on the Big Sur highway are an amalgam of deep-sea sediments like shale, sandstone, and other weakly structured rock. It doesn't take much to perturb these coastal hillsides, so once enough water loosens the soil sitting on the steep slopes comprising minimal support, dirt will move and gravity will do its thing.

On May 20, 2017, a large section of the hillside above the Big Sur Highway collapsed, sending enough dirt and debris downward to cover a quarter-mile stretch of the roadway up to 40 feet deep. Luckily no one was hurt (or killed), and no other damage was reported. Cleanup efforts would not come easy as crews were working on an earlier slide nearby — a constant regularity for this coastal highway.[191]

Instead of trying to clear out all of the debris as they would with most slides, it was decided to just build on top of it, thereby working *with* nature, not against it — something Director Lewin asked the residents of Montecito to do. For Big Sur though this task would be relatively simpler, safer, and not so far-fetched. To prevent future slides, a shelf and ditch were constructed above the roadway to catch falling rocks. Construction though, as of this writing, was still underway and only halfway done, but it is expected to be completed by late summer 2018, at a cost of $40 million.[192] But that won't be the end of it. California's history tells us this is happened before; it can be worse; and it will happen again — and again.

La Conchita

Until the Montecito tragedy of 2018, whenever someone mentioned "mudslide" around Ventura and Santa Barbara counties, one thing usually came to mind: a small, unincorporated community on the coast of Ventura County called La Conchita. With a population of a little over 300, it's a one-light town on U.S. Route 101, so small that if you soak in the coastal scenery while driving past it, you could very well blink and miss it. Yet La Conchita stands as a testament to the burn-rain-slide cycle that can be catastrophically common around Southern California.

With a history dating back to 1817, La Conchita — or Punta Gorda as it was called back then — was a fairly benign coastal hamlet that saw its fair share of farming and other local industries. It also served as a stagecoach stop in the late 1800s and early 1900s, and oil leasing was popular at times around La Conchita as well. A resort hotel was built there in the early 1900s, and in the 1950s and '60s La Conchita became a popular vacation spot for families wanting a quick coastal getaway from the hustle and bustle of Los Angeles to its south. All the while, no major disasters occurred around La Conchita; that is, until 1995.

In the winter of 1994-95 a moderate El Niño was underway. Rain fell at a fairly steady pace in the fall and early winter of 1994, running a moderate three inches or so by the end of December. In January though, a series of very wet Pacific storms brought 13 inches of rain to many parts of Ventura and Santa Barbara counties, and another ten inches of rain fell in March. That's when the hillsides that butt-up against La Conchita could take no more.

Although surface cracks had appeared long before on the La Conchita hillsides, no measures were ever taken to address them. The warning signs were ignored. All the while, as the rains came, water seeped through these cracks, further soaking the soil, and widening the cracks.

On March 4, 1995, the hillside behind La Conchita failed and a mudslide ensued, burying nine homes. Amazingly, no one died. Six days later, another slide would damage an additional five house.[193] It would take another ten years for another slide to occur — but this time with deadly results.

In the winter of 2004-05 another, moderate El Niño brought rain to Southern California. This wasn't your classic El Niño pattern though, with equatorial waters narrowly reaching El Niño status. This rather benign Niño barely budged the jetstream over the North Pacific, and instead draped a leg of the storm-guiding jetstream over the coastal waters off the west coast of the U.S., setting up another potentially wet scenario for SoCal. It almost looked, in fact, like a La Niña pattern with high pressure dominating the Gulf of Alaska. While this usually would result in drier conditions for California — especially SoCal — this can set up a wet pattern if the Gulf high isn't very large, and if it sits far enough out to sea so that its eastside establishes a direct path for Arctic storms to descend southward on a stormy, overwater path to the west coast of the U.S. — all the while picking up Pacific moisture along the way. But during the anemic Niño of 2004-05, a familiar scenario was afoot that would result in yet even more rain.

During December 2004, as the weak El Niño-influenced winter got underway, SoCal saw a decent dowsing with Santa Barbara totaling over six inches of rain by the end of the month. But more rain was coming. As the storm track directed low pressure systems coming out of the Gulf of Alaska southward along the California coast, each storm would eventually encounter an atmospheric river plume of tropical moisture lined up over Central to Southern California. Cold, Arctic storms tapping into this plume would be exceptionally dynamic in nature. Two stormy low pressure systems took advantage of this ideal storm setup with the first acting as a precursor to the second, bringing initially about 3 inches of rain to the Santa Barbara coastal region on the 7th and 8th of January 2005. But the second system was about to drop south from the Gulf of Alaska, this time traversing a longer trek over much more water while increasing in strength.

On January 9, 2005 this second, wetter and more dynamic Pacific storm slammed into the atmospheric river moisture plume while, at the same time, taking an eastward trek toward SoCal as the jetstream bent at the last moment in that direction. Heavy rain ensued that ranged anywhere from 3 to 10 inches over coastal sections of SoCal, and up to an incredible 32 inches in the mountains. Over a 3-day stretch that started on the 7th and lasted through the 10th, Santa Barbara Airport received almost 8 inches of rain. Higher elevations and foothills, like those around La Conchita, saw higher amounts thanks in large part to not just the atmospheric river that had developed, but also from the same kind of orographic enhancement seen right before the Montecito slide, where incoming precipitation piles up against the foothills of the mountains, compacting it into a heavier mass of moisture that squeezes out much more rain. Like Montecito, the storm hitting La Conchita would have similar consequences.

The National Weather Service had issued numerous warnings starting on January 7, 2005 for high winds, heavy rain, and winter storm conditions. These warnings were released to nearly all regions around California. In Santa Barbara County, flash flooding and mudslides closed roads and stranded several vehicles, and mudslides inundated three homes in Lake Casitas. The Ventura Beach RV Resort — which took one heck of a beating during the El Niño of 1997-98 when much of it was washed out to the beach — was flooded, and Highways 1 and 126 were closed due to flooding as well. Across Los Angeles County, flash flooding killed a homeless man in Elysian Park, flooded a mobile home park in Santa Clarita, closed Highway 1, and caused numerous problems in Palmdale. In the mountains, 4 to 12 feet of snowfall was recorded along with southeast winds between 30 and 50 mph. Across the Central Coast and in the Salinas River Valley, high winds gusting to 65 mph knocked down numerous trees and power lines. La Conchita was about to see its worst disaster in history.[194]

On January 10, 2005, approximately 200,000 cubic meters of mud, rock and other debris slid down the hillside backing up against La Conchita. Ten people were killed, 15 homes were destroyed, and another 23 homes were damaged. Overall, damage estimates from the entire series of storms that started in late December 2004 and ended January 10, 2005 were over $200 million.

Real estate is now fairly cheap in La Conchita as getting insurance to cover your property from a slide is next to impossible. For instance, Zillow — the automated real estate value-estimating website — places homes under 2,000 square feet in La Conchita in the $2 million range, yet owners are posting willing bids at a quarter of that.[195] Zillow rates the real estate market in La Conchita as "Cold" — almost no movement for years; however, houses have sold around La Conchita over the past few years, but almost all sold for one-quarter to upwards of one-half of Zillow's value estimates.[196]

The mudslides of La Conchita show the peril of flood and mud around Southern California. While fire caused the Montecito hillsides to lose their soil support structure, La Conchita's hills were compromised long ago, and it didn't take a blaze to do so; instead, easily visible signs of natural hillside deterioration were largely ignored. Yet while there was plenty of evidence ten years prior to the deadly 2005 slide that La Conchita was a prime target for another such event, once again, little to nothing was done to prepare for it. Even as the Thomas Fire bore down on La Conchita in December 2017 — 12 years after the fatal mudslide in 2005 — as flames threatened the tiny town and even reached nearby beaches, some residents of La Conchita defiantly stood their ground, watering down rooftops to protect their beachside homes.[197] The residents of La Conchita, after all these years of repeated and expected disasters, refuse to leave. After all, who can move mountains?

Lessons from Sierra Leone

How we deal with disasters has everything to do with what we learn from the last. Taking a step back each and every time a crisis unfolds, to review what went right or what went wrong, helps us dictate how to move forward. We can see how well governments and agencies used our tax dollars to prepare, mitigate, and recovery from crises. And we as individuals can better decide where to build, live, and prosper. Not all regions around the world though follow the same plans of auditable action, evident when comparing Houston's Harvey to Puerto Rico's Maria. By seeing how *not* to manage the inevitability of nature provides a window where we can view contrasts, which in turn provide us direction for constant improvement.

A prime example of this kind of comparison comes from one of the worst mudslides in recent times, which occurred five months before the Montecito tragedy on the other side of the world. California wasn't the only region experiencing a heavy rainy season after the Niño of 2015-16; Sierra Leone in West Africa experienced a cataclysmic slide in 2017 that was a perfect example of how land-use and alert systems should be optimized, but weren't.

Sierra Leone is accustomed to heavy rainfall, surrounded by heavily forested mountain ranges that, like the mountains and hillsides around Montecito and La Conchita, exacerbate rainfall amounts by collecting moisture from the orographic enhancement mentioned earlier. Starting in July 2017, Sierra Leone was enduring an exceptionally rainy season that brought an unthinkable 41 inches of rain to the area in just a matter of weeks. Water management isn't Sierra Leone's strongest suit, suffering from a long-term issue of poor urban development and unplanned sprawl. Officials failed to stop housing developments that encroached on local flood plains, and hodge-podge municipal works resulted in much narrower water passageways than the area should have been allowed.[198] Negligent waste removal left drains to fill with sewage and debris, as finger-pointing between local officials distracted from the problem, resulting in no efforts to resolve the issue — and subsequently no escape for rushing waters flowing off the nearby mountainous regions.[199]

Further compounding Sierra Leone's problems was deforestation that weakened hillsides on its surrounding mountains. As we know from La Conchita's history, fire isn't the only factor that can weaken a slide-prone hillside; nature — and especially the mismanagement of it — can too. In the case of Sierra Leone, humankind was to blame for putting hillsides in a perilous position.

After nearly tripling the area's seasonal rainfall average by August, with more heavy rains on the way, Sierra Leone's meteorological department failed to issue any warnings about rising flood waters and the potential for slides. As residents continued their daily routines, accustomed to wet, West African rainy seasons, they were left unaware that the hillsides around them were weakening at alarming rates. On the night of August 13, 2017, Sierra Leone residents went to bed listening to the sound of pouring rain. Many didn't realize it would be their final moments.

As a period of non-stop, torrential rain swept through Sierra Leone on August 14, 2017, Sugar Loaf mountain — overlooking Sierra Leone's capital of Freetown housing a population of more than 7 million — endured a partial collapse, triggering mudslides that buried numerous homes, and residents as they slept.

Confusion ensued as the slide further affected a dozen more settlements in the region. Accessibility between communities was cutoff as bridges and roadways around the area sustained damage. Power outages left millions without electricity as the local utility company temporarily cut power to communities as a precautionary measure. When it finally came to an end and the mud stopped flowing, more than 1,100 people were declared dead or missing, and there was over $15 million (US) in damages.[200]

These tallies would have certainly been lower if the Sierra Leone region had better managed their surrounding environment. While homes in Ventura were constructed with less than perfect fire-protective roofs, and perhaps many properties affected by the Thomas Fire were too close to desiccated open-space, land management and development zoning in the U.S. is far better than it is in many foreign lands, like Sierra Leone. However, no matter how well we plan our developments, and no matter how well we build our homes and communities, one of the biggest problems we see time and again — with the Thomas Fire, Montecito mudslides, Hurricane Harvey, and Sierra Leone as well — is letting people know they need to either get out, or die. Communication is key to saving lives, but sometimes that breaks down, and fails. Ending Act 4 of our story, this is where the trail leads us next, covering something that affects all elements of nature's wrath equally.

Communication Breakdown

Uh oh, what's going on?

~Director of the emergency management office in Santa Barbara, Robert Lewin, when, after issuing an emergency alert, saw that only a small number of cellphones were alerted of impending mudslides in Montecito on January 9, 2018, leaving residents unaware that some would soon be buried alive in a deluge of mud.[201]

In early 2010 I was hired to work on a new project by NOAA's National Weather Service (NWS), where I had the chance to see firsthand how government bureaucracy can utterly waste taxpayer money. Hired as a Senior Software Engineer for a government contractor in Camarillo, California, funded by NOAA through the Department of Commerce, I was thrilled to work with brilliant-minded meteorologists, yet frustrated and disappointed by the incompetence of many who managed them. Although intentions were grand by government officials and employees, their inability to get things done would put people's lives at risk. Embarking on what seemed like a rewarding project that could help make the world a better and safer place, my enthusiasm turned to weariness and disappointment over the next couple of years.

Shortly after Hurricane Katrina in 2005, NOAA, funded by the Department of Commerce, worked on a proposal to improve their outdated weather radio system. In case you've never heard or seen such a thing, these radios, which have been around for decades, would sit in homes, classrooms, and offices usually unnoticed, until a tornado, hurricane, fire, or other disaster alert was broadcast, similar to how those emergency broadcast messages sometimes appear on your television screen. The idea for this improvement project was a noble, yet overly lofty. Instead of just upgrading the aging radio system's broadcast infrastructure that had become so old and deprecated that replacement parts were no longer available, NOAA would not just update the system, but at the same time expand it to allow integration from the Department of Homeland Security (DHS), the Federal Emergency Management Agency (FEMA), the National Law Enforcement Telecommunications System (NLETS), and potentially other emergency management agencies as well. By having all of these agencies working together to disseminate local and national emergency voice alerts, it was thought that much of the Tower of Babel-like debacles during the Katrina disaster could be avoided. It was a great idea, but the federal government, along with its various agencies, simply could not get their acts together. The White House administration at the time, under then President Obama, would make things worse. But this wasn't a partisan problem.

During the 2005 Katrina disaster there was more than enough finger-pointing and criticism to go around. New Orleans's Mayor Ray Nagin was criticized for failing to implement a food plan while ordering residents to shelters, which led to the squalid conditions in the Louisiana Superdome that later became a site of rapes, drug dealing, vandalism, violent assaults, and gang activity.[202] Even more blame was aimed at the Bush administration for apparently dropping the ball with a glacial-paced emergency response. New Orleans's emergency operations chief Terry Ebbert blamed the inadequate response on FEMA, stating,

> *This is not a FEMA operation. I haven't seen a single FEMA guy. FEMA has been here three days, yet there is no command and control. We can send massive amounts of aid to tsunami victims, but we can't bail out the city of New Orleans.*

History's political blame-game repeats itself, just like the tit-for-tat Twitter rampage between President Trump and San Juan's mayor Carmen Yulin during the aftermath of Hurricane

Maria in 2017. After a disaster hits, life will not return to normal quickly, yet frustration by the hurt and hungry will inevitably lead to ridicule and criticism of those in charge. Communication — it was thought in 2005 by the NOAA project proposers — was the culprit, so fixing that would solve the world's disaster woes. Which made sense, at the time.

Mayor Nagin also expressed his frustration during Katrina disaster-relief efforts at what he claimed were insufficient reinforcements provided by President Bush and federal authorities. At the time, the Department of Homeland Security was only three years old, and officials there were still trying to get their footing after the 9/11 attacks that spawned the formation of that organization. With FEMA flailing as well, there appeared to be a need to tie all of these loose ends together so that right hands knew what their lefts were doing.

Thus, the crux of the Katrina problem was seen as a basic breakdown of communication, and the people at NOAA wanted to fix it. Designing a multi-agency alert system seemed like a good idea at the time. But nothing comes easy when it comes to the feds, and the world would change faster than Uncle Sam could keep up.

It took almost five years for funding to be approved for this new NOAA, multi-agency project, called the "Weather Radio Improvement Project", or WRIP for short.[203] Those five years are an eternity on technology timescales. Uber, by comparison was funded in four months — not years, months — proving that private enterprise can move about 92% faster than our federal government.[204] And as should be expected, during this half-decade of legislative wrangling, technology was rapidly changing, eventually putting the idea of a weather radio into the archives of techno-past.

By 2010, when the WRIP project finally entered into its design phase, more people had cellphones, Twitter, Facebook, text messaging, and other more modern means of communication that surpassed a portable weather radio you could pick up at your local Radio Shack store. But that didn't deter the powers that be from plowing forward while stepping backward in time.

Once WRIP was funded, it took another year just to come up with a design that would pass the plethora of approvals. Progress ran at a glacial government pace as numerous cooks in the proverbial kitchen wanted to not only have their say, but also make sure that their visits to California for design and review meetings were justified by their agency's budget. There were actually some government representatives who had the gall to, after meetings, check with the meeting-minutes taker to ensure they included their comments; after all, their bosses were watching, and they wanted to make sure they were relevant, and that their trip was worth it. Forget the fact that in meetings holding dozens of government agency representatives, some would fall asleep, seemingly unfazed by who would notice.[205]

With all fairness to the National Weather Service, their IT department was fairly impressive — not astounding, but good nonetheless. The IT guys had stated in their initial proposals that they wanted to use open source components and modern day computer languages to build the new system — things that are used today in other high security systems. Yet the contractor who employed me for the project thought it best to use their own inventive yet outdated software and systems, and forced this upon the NWS's IT staff, despite their disapproval.[206] It were as though I was writing software in the dark ages of computing in a slow-moving time-warp powered by the gravitational-pull of a dense, bureaucratic black-hole. As the project's development progress churned at ever-slower speeds over the next couple of years, post-Katrina intentions, as good as they may have been, were becoming more burden than blessing. After a couple of years, we'd be put out of our miseries.

Halfway through 2012 the WRIP project was put on hold — not by the ongoing snails-pace of technology progress or snoozing agency representatives, but by budget cuts enacted by the Obama Administration. Asking for an additional $163 million to be spent on weather satellite programs would comprise about half of the entire NWS budget, leaving nothing left over for hardly anything else.[207] To accommodate this increase, cuts had to made elsewhere, from top to bottom. WRIP got the axe, as did employees in the National Weather Service, which is a problem that continues today, but is rarely discussed, or made public. While ending a project like WRIP could be questionable, nixing jobs of the people watching life-threatening storms is not.

A tally from the National Weather Service Employees Organization — the union representing NWS employees — shows a steady decrease in staff since the Obama-initiated, WRIP-killing budget cuts from 3,802 non-managerial workers to 3,368 — a decrease of more than 11 percent. Some of that is also attributed to hiring freezes over the past few years, as well as an attrition problem as wary employees, looking for job security elsewhere, abandoned ship.[208] And who can blame them? After all, more cuts are on the chopping block as I write these words.[209]

To put your trust in taxpayer-funded, government agencies to protect the population would be folly. But that's what we've relied on since 1870 when a joint congressional resolution required the Secretary of War (William W. Belknap) to,

> *...provide for taking meteorological observations at the military stations in the interior of the continent, and at other points in the States and Territories...and for giving notice on the northern lakes and on the seacoast, by magnetic telegraph and marine signals, of the approach and force of storms.*[210]

The NWS has improved their forecasting since then, and has helped save the lives of countless people, warning of impending disasters. But as *communication* technology supplants weather radios, and the NWS's websites lack much of the luster that other, commercial weather services provide, the NWS's constant budget cuts threaten the lives of many more. A roof doesn't leak until it rains, and you don't realize how important the NWS is until you need it.

As the WRIP project chugged along toward its eventual swansong, a more modern emergency alert system was getting underway. The Wireless Emergency Alerts system (WEA) was being developed in 2008 and was released when WRIP died in 2012.[211] Unlike WRIP, the WEA system has the potential to save many more lives as it is intended to broadcast public safety alerts to your cellphone using FEMA's Integrated Public Alert and Warning System (IPAWS). The NWS has a direct link to this alert system via the Commercial Mobile Alert System (CMAS), where they can push notifications concerning tornados, flash floods, dust storms, hurricanes, typhoons, extreme wind, thunderstorm warnings, and tsunami warnings as well.

This same system was used for the false-alarm of an inbound missile to Hawaii on January 13, 2018, when residents went running in panic, thinking a nuclear warhead was headed their way from North Korea. While the system worked well, the person sending the alarm had a history of confusing drills and real-world events.[212] Not knowing that a drill was underway, the now unemployed worker triggered the real deal, sending an alarm to cellphones across Hawaii that said,

> *EMERGENCY ALERT: BALLISTIC MISSILE THREAT INBOUND TO HAWAII. SEEK IMMEDIATE SHELTER, THIS IS NOT A DRILL.*

But, it was.

The Thomas Fire also encountered false alarms from the WEA system. On Sunday December 10, 2017, as the Thomas Fire was raging in full force, numerous Ventura County residents were woken up at 2:19 AM with a Civil Emergency notification to "Evacuate Now." It was a mistake, which was corrected about a half hour later, clarifying that the order was only for the area north of the city of Carpentaria. The director of the emergency management office in Santa Barbara, Robert Lewin — who I mentioned earlier having similar troubles with the Montecito mudslide alert — took the blame for the Thomas Fire false alarm, saying:[213]

It's a complicated system. One of the boxes was inappropriately checked, despite our training and having our very best person sitting right next to me preforming it.

Because of one inadvertently clicked checkbox on a government-designed, complex system that is intended to do one simple thing, a false-alarm was sent to thousands.

Like any technology, the new WEA system has some snags. First, you may not realize that your cellphone has this feature enabled. On an iPhone, for instance, you can go to Settings, then Notifications, and scroll all the way down until you see "Government Alerts". From there you can select "Emergency Alerts".[214] In many cases, you might find that it is disabled. Enable it now.

The WEA system also can't pinpoint your exact location all the time. Relying on cell towers — that your phone connects to — could be inoperable from a disaster, so there is the possibility that an alert may not reach you, thinking you are far away from the disaster at hand.

But the biggest problem with receiving these wireless alerts to your phone is simply the people sending them. To complicate things further, it's harder for those sending the alerts to know who all will receive them. When discussing the Thomas Fire false alarm, Director Lewin went on to say:

We're leaders on this system, but it's extremely complex, and it needs some refinements. For example, WEA is unable to deliver an evacuation order with surgical precision, and it's limited to 90 characters and doesn't have a map feature. We are hoping to partner with federal and state government for a system that's easier to use...We're trying to get absolutely accurate information out there and we're dealing with multiple agencies. Mistakes have been made, and we try to correct them as fast as possible."

Since then, starting in February 2018, the Santa Barbara County Office of Emergency Management enacted a new, and more complicated timeline to issue evacuation orders that start as early as 72 hours in advance, beginning with a "Pre-Evacuation Advisory"; 24 hours later a "Recommended Evacuation Warning" is issued; 24 hours after that a "Mandatory Evacuation Order" is issued; and then, finally, 12 hours later, the mandatory order becomes effective.[215] But using a more involved, process-centric approach to solve the problems of past notification meltdowns is reminiscent of the complicated color structure that was used — but since been replaced — to gauge terrorist threat levels.[216] The food pyramid is less confusing. Instead of making new color-coded flow charts and diagrams with a wider array of terminology for people to wrap their heads around, it'd likely be best to just let people know they need to get out. Having more links in the chain leading to evacuation events provides additional points for potential failure. Besides, charts and timelines weren't the crux of the communication problems; technology and the people using it were.

But with all fairness to Director Lewin, when you work within a slow, lumbering, bureaucratic machine like the state, local, or federal government, there is only so much you can do. If your hammer is managing process, then charts and graphs are your nails. Designing technology systems for disseminating alerts is not typically within the purview of an emergency management director, since other cogs comprising the government machine have that responsibility. So if it seems like I'm picking on Director Lewin throughout this book, rest assured that is not my intention. Mistakes were certainly made on his part, but a much larger onus goes higher up the chain, and across various departments as well. No time during my tenure working on the NOAA weather radio redo project do I ever recall meeting with any emergency management director at a county level, even though the new system was supposed to help these local offices get the word out. That's not the fault of Director Lewin or his contemporaries;

instead, it's a blind-spot in the overly complex workings within the federal government that provided funding and subsequent oversight for projects designed (with intent, at least) to improve disparate lines of communication. It's an inherent, governmental catch-22 kind of paradox, where action degrades intent.

While Lewin and his team train to use the newer timeline, and they continue to struggle with the new alert system, WEA has become, in many ways, better in today's world of technology and communication than the weather radio project of the past. But, it still requires someone to issue an alert, leaving some degree of subjectivity to whether an alert should be sent at all, which is a problem.

During the Northern California fires of 2017, which resulted in 44 deaths, local officials in Sonoma County chose not to issue any alerts. Their odd justification was that they feared pandemonium and clogged roadways would make it difficult for first responders to get to people in harm's way, yet people wouldn't have been in harm's way if they would have evacuated. That paradoxical and nonsensical decision left numerous Sonoma County residents with no official warning. In fact, many fled only after being awakened by the smell of smoke, the sound of sirens or neighbors pounding on their doors. It's the same bad judgement noted earlier in the "Flood" chapter when Houston's mayor never gave the order to get out of the way of Harvey, thinking that there would be an evacuation "nightmare".

One has to now ponder another kind of paradox: Why do we have alert systems if the people in charge won't use them; or why do we have those particular people in charge? The latter seems more pertinent than the former.

Sonoma County officials did though send text messages to one of their local programs that you would have had to sign up for, and they placed robocalls to landline telephones — a further sign of how out of touch these officials were with the people they were there to protect.[217]

It's great to see that communication technology has advanced to the point where we can get a text message disaster alert on mobile devices that are virtually leashed to each and every one of us. But technology only works if it is utilized. In the day and age of weather radio alerts, your weather radio could sit in a silent, standby mode until an alert came in, which would then sound the alarm and tell you of the danger ahead. The determination to turn on a dormant weather radio was from a simple header embedded in the message, added there by the direction of someone at the National Weather Service who was issuing the alert. But since the new WEA system allows numerous agencies to send alerts, things are a bit more complicated, and as Directory Lewin correctly points out, can easily lead to mistakes from an overly complex, poorly designed system that has severe limitations.

FEMA has a fact sheet[218] that's given out to users of the WEA system, giving them guidance on issuing an alert. FEMA's recommendations (not requirements) to decide whether to issue a public alert or warning has three seemingly simple points to follow:

1. Does the hazardous situation require the public to take immediate action?

2. Does the hazardous situation pose a serious threat to life and/or property?

3. Is there a high degree of probability the hazard situation will occur?

Nowhere in this WEA charter does it say anything about concern for potential pandemonium from people trying to evacuate — a justification used time and again for not alerting the public about the Sonoma fires, Hurricane Harvey, and at one point, the Montecito slide.

The three-page guidance fact-sheet from FEMA goes on to tell WEA alert-issuers how to form effective warning messages so that actions can be taken by the public to protect themselves. These "Components of Effective Warning Messages" include:

- Specific Hazard: What is/are the hazards that are threatening? What are the potential risks for the community?

- Location: Where will the impacts occur? Is the location described so those without local knowledge can understand their risk?

- Timeframes: When will it arrive at various locations? How long will the impacts last? When should people take action?

- Source of Warning: Who is issuing the warning? Is it an official source with public credibility?

- Magnitude: A description of the expected impact. How bad is it likely to get?

- Likelihood: The probability of occurrence of the impact.

- Protective Behavior: What protective actions should people take and when? If evacuation is called for, where should people go and what should they take with them?

All of which would make sense if you could fit all of that into the 90-character limit of a WEA message.[219] You read that correctly: 90 characters is the maximum limit for a WEA message. When your life is in danger, and there are mere moments to spare, someone has to come up with a creative headline-length message that will give you all of the appropriate and critically important information to save your life. Twitter has had nearly twice as much space for utilizing such a message, and the new 280-character Twitter limit being rolled out puts a WEA message to shame. Yes, size *does* matter, especially when talking about an alert that makes the difference between life or death.

FEMA's out of touch guidance fact-sheet for the 90-character limited WEA messages goes further to recommend that issuers need to avoid "poorly written warnings" that can "undermine both understanding and credibility." To ensure your 90-character message is well written, FEMA recommends the messages should be:

- Specific: If the message is not specific enough about the "Who? What? When? Where? Why? How? the public will spend more time seeking specific information to confirm the risk. Be specific about what is or is not known about the hazard.

- Consistent: An alert/warning should be consistent with messages that are distributed via other channels. To the extent possible, alerts/warnings should be consistent from event to event, to the degree that the hazard is similar.

- Certain: Use authoritative language and avoid conveying a sense of uncertainty, either in content or in tone. Confine the message to what is known, or if necessary, describe what is unknown in certain terms. Do not guess or speculate.

- Clear: Use common words that can easily be understood. Do not use technical terminology or jargon. If protective instructions are precautionary, state so clearly. If the probability of occurrence of the hazard event is less than 100 percent, try to convey in simple terms what the likelihood is of it occurring.

- Accurate: Do not overstate or understate the facts. Do not omit important information. Convey respect for the intelligence and judgement of your public.

Ironically though, I had to correct six spelling errors in this list when I added it above. Although FEMA recommends "Effective Style Guidelines" for warning messages so they are not "poorly written" to "undermine credibility", they don't seem to use spell-check on their own public documents. And to add insult to injury, FEMA acknowledges that these alerts and warnings will be accessed by persons with disabilities using text-to-speech conversion software, and also by non-English speaking people as well. Hopefully such software can accommodate FEMA's spelling errors.

When you think about it, emergency management officials in the United States using the new WEA system have a lot on their plate, which is further complicated by the frustrations of shoving enough information into a 90-character message using a non-user-friendly interface. Like the WRIP project before it WEA, though well intentioned, shows how technology designed by the hands of the federal government is often out of touch — which is why the feds often out-source their work to contractors like Lockheed Martin, Boeing, General Dynamics, etc. So should our emergency management agencies be farmed out as well? That may not be in the best interest of *we the people*, but it's come under consideration more than once.

Each year AccuWeather, Inc. — a commercial, for-profit weather company — spends tens of thousands of dollars lobbying the House, Senate, and Commerce Department on what they deem as "commercial weather industry issues."[220] That's esoteric politico-speak for wanting to privatize the nation's weather information into for-profit companies, like AccuWeather. AccuWeather has upped their game recently, from spending $22,000 on lobbying in 2001 to $90,000 per year in 2016 and 2017.[221] Although AccuWeather gathers and republishes data provided by the National Weather Service — which you pay for with your taxes — AccuWeather would like for you to pay for it again by watching and/or clicking on their website ads or other means of monetization that they may come up with. Trump was initially on board with this idea when he tapped Barry Myers, the former CEO of AccuWeather to head-up NOAA. On December 13, 2017, the Senate Commerce Committee voted to move forward with the nomination of Myers for that position. Given the lobbying track record of AccuWeather, this could be disturbing, or it might be a way to right a tilted weather ship.

Things can go wrong (or sometimes right) when non-profit organizations become commercialized. Blue Cross Blue Shield (BCBS) is one such example, which falters on a fine line between non- and for-profit status, leading to some of its state tax-exempt statuses being revoked in 2014 after it was found they were holding onto $4.2 billion (yes, with a "B") in its financial reserves — an excess for a non-profit of its size.[222] Rising health care coverage (not medical costs) have been impetus to numerous reform attempts by Democrats and Republicans alike. It's a broken system exploited by "wolves in sheep's clothing" for-profit companies like BCBS, who are largely to blame.

Then again, perhaps putting a for-profit weather company in charge of our nation's weather may result in better designed alert systems that compete with every-day, free communication systems like Twitter that allow more than 90 characters to be transmitted at a time. Perhaps putting competition on the playing-field will spark further innovations with weather and alert systems that go beyond what our imaginations can dream up at this time. And perhaps having competitive companies being in charge of disseminating weather information will result in better-read documents void of spelling errors and false missile alerts. Moreover, perhaps a company run like any other commercial, non-governmental agency, will keep people in line so that they won't hesitate to make critical decisions that should put people first, and the fears of pandemonium last. Perhaps this could save more lives. But it could come at a cost.

Commercial products survive through competition, which oftentimes leads to heterogeneous, disparate systems. The VHS and Betamax systems in the old days of video-tape technology are prime examples of differences in competing systems, requiring people to use one or the other, but with no compatibility between them. Apple iPhones versus Google Pixel or Samsung Androids are modern day dilemmas that spark conflict when working toward a common goal — like a non-partisan, non-profit, WEA system. Imagine an incredibly awesome alert system made by a for-profit company, but it requires you to install their app. Compatibility between various mobile manufacturers would add competitive-complexity to the mix, thus being a less effective system.

Additionally, having a private, commercial company controlling life-saving information could lead to further communication disruptions if an up-and-comer startup tries to supplant the incumbent technology — which always happens in the commercial world of tech. And in a country where monopolies are discouraged but companies have autonomy and freedom, taxpayer-funded government agencies would have little say in how technology is designed, built, or used to get critical information to its citizens.

Perhaps we should just try to fix the alert system we have now. But things like this don't get attention until they are used, and fail.

But one has to consider a bigger, more overarching question when it comes to disaster communication: Why are federal and state governments willing to spend tens or hundreds of billions of dollars on natural disaster recovery, yet the budget of the National Weather Service, which is a mere fraction of that, continues to be cut? With the NWS budget at $1.1 billion per year, it would take merely one-tenth the cost of Hurricane Maria's recovery efforts to double what the NWS has to work with. It's a drop in the bucket that could have a much more positive impact — a high return on a small investment. Taking a proactive measure to double the National Weather Service's yearly budget could put more eyes on the seas and skies, and perhaps also boots on the ground and alerts to your phone. Yet somehow it's become acceptable to dole out a hundred billion or more dollars of taxpayer money toward recover from a disaster than to spend a fraction of that on preventive measures that could reduce those costs, and save more lives. Somehow though politics get in the way — they did during the Obama administration with WRIP and NWS cutbacks, and they continue in the current Trump administration as well. It's not a party problem; it's a systemic political dilemma that spans both sides of the congressional aisle.

I admit it's easy to lay blame on those who made mistakes. And I know firsthand that it's hard to come up with alternatives to systems and processes that may be inefficient, or just don't work. While I may have made emergency managers in this book look like Keystone Cops, I will admit that, after working on an emergency alert system myself, I don't have all the answers. But it seems it would make sense to invest more into systems that keep us safe rather than aftermath recovery efforts for the disasters that, although not preventable, could have been better mitigated by well-funded, better staffed and managed systems. But as Weather Service budgets continue to be cut by presidential administrations (from both sides of the aisle), and lobbyists vie for privatization with intentions aimed for their own self- and corporate-interests, disasters will come, and disasters will go, in a flow that puts all of us in harm's way.

Repairing breakdowns in communication could not only save lives, but taxpayer money as well. Yet for some reason, no matter who or what party is put in charge to captain our country's ship, simple life-saving steps continually get overshadowed by insurmountable walls of bureaucracy, a blindness to foresight, and downright negligence. Federal and state budgets, for some reason, seem to ignore even the slightest hint of proactive disaster measures, while at the same time, the citizens who fund these governments face peril, injury, and too often, death.

Epilogue

Life is like riding a bicycle.
To keep your balance,
you must keep moving.

~Albert Einstein

Things are not always what they seem, or what we wished they would be. As technology advances we often drift away from utopian dreams of a futuristic world at peace and harmony with nature and humankind. Instead we're drawn to the loudest cries, the most shocking scenes, and the brightest shiny bobbles placed before our world of non-stop connection, communication, and media. While weather models improve and science unravels meteorological mysteries, their mundaneness competes with interests far from the natural world, guiding our attention elsewhere, and often inappropriately.

An example of distraction happened in 2017 when the media was whipped into a frenzy after the Trump administration allowed the charter for the "Advisory Committee for the Sustained National Climate Assessment" to expire. This climate advisory committee would not be funded further, and would essentially be disbanded. Shock-and-awe headlines followed:

> *"The Trump administration just disbanded a federal advisory committee on climate change"*
>
> *– The Washington Post*

"U.S. Disbands Group That Prepared Cities for Climate Shocks"

—Bloomberg Politics

"Trump closes panel meant to help cities deal with climate change"

— The Hill

"The Trump Administration Really Has Very Little Use for Scientists"

— Mother Jones

"Exiled by Trump, climate scientists lead the resistance against the denier-in-chief"

— The Independent

...among others.

It appeared as though science was thrown by the wayside, abandoned by malevolent non-believers who cared more for their bottom-line than the health of the planet. Not only was this perception false, it was also a serious distraction from more critical issues at hand.

During the first four months of 2017 — as embellished headlines cried foul on climate control efforts — the United States endured more than 5,300 severe weather events including tornadoes, large hail, and wind damage. This series of weather events was more than double the average.[223] But you may never have known that unless your *specific* town was affected. The few weather stories that did get attention in 2017 were Harvey, Irma, and Maria — a small fraction of what occurred weather-wise in the United States. Nevertheless, the headlines about a governmental climate breakup took the forefront, and were shocking — for more than one reason.

Unless you could get past the hyperbolic headlines, stock footage of disastrous floods looping over and over again, numerous paragraphs preceding the fine-print, and the plethora of pundits touting either disaster or denial, you might have missed something. The Trump administration was not disbanding anyone doing climate research; instead, this advisory panel had a much simpler charter that was, in many regards, a complete waste of taxpayer dollars. Before you click "Send" on that hate-mail you're constructing to send my way, please allow me to explain first.

The "Advisory Committee for the Sustained National Climate Assessment" that was disbanded by the Trump administration in 2017, was put together just a couple years prior, in 2015, to basically decipher climate reports that were found to be too complicated for public officials to read and comprehend. That's right: a taxpayer-funded advisory panel was put together and paid for through the U.S. government because they found climate reports to be overly complex; so much so that they needed to hire 15 people for this advisory board to translate the reports for them. The report this particular team of 15 would look at was generated each year by not the IPCC, but by yet *another* government outfit called the U.S. Global Change Research Program (USGCRP), which comprises a team of 300 experts, guided by still yet another 60-member Federal Advisory Committee, which has been around since 1989.[224]

Let's step back and chew on that for a second: A 300-member government-funded program that started in 1989 produced climate reports so complicated that it required yet another advisory committee of 15 members to simplify those reports for politicians and other government officials starting in 2015. This, despite the fact, that the USGCRP, like the IPCC, provides simplified summaries with easy-to-understand, high school-level reading. Why, after 26 years of the USGCRP and Federal Advisory Committee, did the Obama administration find it necessary to create this 15-member translation team? Why didn't someone in the Obama administration just simply ask the USGCRP to give them a simple synopsis, like the IPCC does for policymakers? The answer to that would have been quick, since, as a matter of fact, they do. Each of the USGCRP reports has an "Executive Summary" that precedes each of their reports. But for some reason it was thought that this was still too complicated. Trust me, it's not; you can read it for yourself; just follow this endnote.[225]

The additional 15-member advisory committee that was to advise information from another advisory committee was unnecessary, and redundant. Welcome to the United States government.

Aside from that head-scratcher, how did government officials get elected or hired without a basic understanding of climate change? It's not like this is new stuff here — Al Gore gave everyone a running start with a primer on the subject starting in 2004. He even made a movie about it — you didn't even have to read, just grab your popcorn and watch. Cursory knowledge of climate change should be a prerequisite for any government official, especially since they like to refer to global warming at press conferences covering disasters, using climate change as a scapegoat in a line of questioning that, although overly simplified, goes something like this:

Reporter: *Why did this happen?*

Government Official: *Climate change.*

Reporter: *What will you do to prevent this?*

Government Official: *The climate is changing and we're all to blame. End of story.*

The distraction comes full circle as government officials prefer to turn the tables, placing the blame on everyone (and everything) but themselves. The media eats it up, and then spits it back out into climate-grabbing headlines. Politics and bias further accentuate the problem.

Distractions are more than an annoyance; they are roadblocks standing in the way of progress. Officials all too often fail to put forth meaningful plans to mitigate the disasters from the climate change they preach — they only tend to focus on centuries-long effects. Such line of questioning that ends with buck-passing answers is seen time and again, like Governor Brown's "little green lawn" comments noted in the "Flood" chapter; or worse, like when UN Secretary-General, Antonio Guterres shifted the line of questioning concerning Hurricane Irma (noted in the "Flood" chapter as well). This is a blatant abdication of responsibility. By placing the cause of disasters solely on climate change, officials shift the blame away from their purview, inaction, and in many cases, incompetence. It's the ultimate distraction — a political sleight of hand.

California, which often sets an example for environmental awareness, has a climate change program that's been in place since 2004, but has barely been touched for the past three years. Even if it was, it's mostly just a collection of ineffective, bureaucratic policies enacted by Governor Brown to fall under his "Five Pillars" plan, which was designed to combat global warming in the state.[226] The lion's share of Brown's pillars plan concentrates on reducing greenhouse gas emissions, and taking some steps to conserve water (by mostly shutting off your tap). Governor Brown, following this plan, signed a series of executive orders — a way to bypass state legislature that would normally vote on such issues — from 2004-15 requiring reductions in greenhouse gas emissions, setting up various "green buildings", and playing musical chairs with agency responsibilities to put more cooks in the climate-watch kitchen.[227] But no steps have been taken to actually address how to deal with disasters that could come from a warming world. Instead that, like Oroville, is always an afterthought.

Governor Brown's pillars plan on climate change is a prime example of how California's leadership tends to hug more trees than people. Once a disaster strikes though, state governors, like Brown, are quick with the executive pen to enact immediate relief efforts and start signing checks — which is a good thing. Yet at the same time, these officials hold press conferences publicizing their good deeds, while not taking responsibility for failing to protect the people they are now trying to save. Instead, it's easier (and far more popular) to just direct everyone's attention toward climate change, and just go back to business as usual.

Not once did California's state government address potential dam failures, like Oroville, in an actual climate change action plan. Environmental advocacy groups filed a motion in 2005 over concerns regarding the Oroville dam's condition, but their warnings fell on the deaf ears of the Federal Energy Regulatory Commission. And while upgrade proposals were rejected, including $100 million by community groups around Oroville, nothing was done. Governor Brown and his team of Five Pillar climate watchers took no action either.[228] It was only until Oroville failed did Governor Brown take any kind of action, by signing an inspection bill for Oroville. To add further insult to injury, California's state legislature voted to keep dam-safety plans secret, for what they say is to "protect public safety" — by not letting us know how they are spending our taxes.[229]

Al Gore has sailed a similar tack on his sojourn of climate conquest with his non-profit organization, "The Climate Reality Project", which serves mainly as a means to promote and sell his climate books and films. Their mission statement is:

> *To catalyze a global solution to the climate crisis by making urgent action a necessity across every level of society.*[230]

Ambiguous as that statement is, Gore's group goes on to mix fear with the idea that talk can produce intended results, while insinuating that pulling out of the Paris Agreement was a mistake on the part of President Trump. Acting more like a criticizing cheerleader than a helpful coach, Gore's Climate Reality Project mission statement crescendos with a pat on his back by saying:

> *We're working to accelerate the global shift from the dirty fossil fuels driving climate change to renewables*

so we can power our lives and economies without destroying our planet. But we can only do it together. That's why we're uniting millions from all corners of the globe and all walks of life — factory workers and farmers, teachers and taxi drivers, students and shop clerks, and more — to make our leaders honor and strengthen their commitments to climate action that they made in the Paris Agreement. Stopping climate change is the challenge of our time. But we know how and thanks to the Paris Agreement, we have the tools to do it. The sustainable future we want is finally in our hands. And at Climate Reality, we won't let it slip away.

Which all sounds fine and good. But in reality, all that Gore's non-profit group has done — and continues to do — is preach. They've done little else. There are no fundraisers for hurricane victims, or redesign efforts to avert future disasters like Maria, Harvey, Irma, Oroville, etc. When Montecito went down, Gore was nowhere to be found. Of the $26 million that The Climate Reality Project raised in 2015, none went to victims of the disasters that Gore blamed climate change on. The President and CEO of Gore's non-profit though, Kenneth Berlin, was given a salary of $309,517 in 2015.[231] And Gore himself has raised his net worth from $1.7 million from when he was vice president to now $200 million — a great deal of which has come from riding the climate change campaign train.[232] Gore has become richer, yet he has taken no actionable steps to save lives from the perils he preaches.

Much attention goes toward press conferences by government officials, and celebrities like Gore to preach the carbon sins of a warming world. It was a natural segue then to show more media coverage on President Trump's disbandment of the redundant advisory council than the lack of attention given to preparing for climate change disasters. It's an ironic abdication that plays out too often: Complain about climate change but take no steps to prepare for what it could bring. Many of our elected officials set that example, following in the footsteps of Al Gore.

I don't care who you voted for, what political party you align with, or what your feelings are about climate change. The common ground I'm sure we can all agree on is that climate and weather are not just media rants and ravings for improved network ratings, but also fodder for political agendas, and in some cases, profitability. The headline hyperbole that went hog-wild in 2017 regarding the short-lived, 15-member, redundant and essentially useless advisory panel set the tone, and was the epitome of distraction examples.

Information is imperative to survival, and preparation is paramount. While weather events may quietly escape our memories over time, causality linked to them is sometimes used subjectively as a means of confirmation bias. Cherry-picking pieces of data while not understanding the full story is how myths are made and dogmas flourish. When things that are preached never reach fruition, non-believers and the undecided often turn a blind eye. While some may feel that over-hyping a cause like climate change is warranted to make everyone wake up and take action, it actually muddies the waters for long-term problems that are hard for human short-term memories to fully grasp.

Even during my final words to end this story I've exhausted an effort to rationalize a call to reason. Yet this, in some ways, like overly hyped headlines, is an unnecessary distraction too. Instead of defending opinion or dissecting here-today-gone-tomorrow news-cycles, I believe our concentration should lie elsewhere.

While opposing sides of any debate continue their disagreements or turn away from the constant cries of false prophets publicizing misleading headlines, we should remember the victims of the tragedies that have unfolded over the past few years — and beyond. Loving families were torn apart; once-laughing children were buried; and beloved pets that often bring out the best in our humanity were burned alive. Heroes fought uphill battles against the worst of nature's fury to save property and lives. And many who survived lost all that they worked for. By remembering the human cost from nature's tragedies, perhaps we can improve our own humanity as well, thus further motivating us to make this world a better place. Our forebears' progressions provide what we have today, but the work to improve is always incomplete. We must continue to march forward — undistracted, and undeterred.

We've learned a lot since the Great Flood of 1862. We've improved zoning and building codes, but it's an on-going process that's never finished. Looking back at the Oroville crisis, Houston's flood plains and dams, Puerto Rico's easily collapsible infrastructure, and the less than adequate rebuilding codes for recently fire-ravaged Ventura, it's evident we need to do better on choosing not just *how* but also *where* we build, and with *what*. Knowing the eventual inevitability of where water will flow, winds will blow, fires will burn, and mud will slide, should set ever-higher standards through a never-ending process of continuous integration, turning what we learn into better planning and preparation. Rebuilding merely repeats history with a blind eye to the future. Improving how that is done though will continue the momentum of human progress.

Since 1862 we've gotten better at forecasting weather as well. We've improved our understanding of climate cycles, and how they impact seasonal changes. When El Niño comes, some dust-off their bigger boards, while others prepare for rain. Once the surf settles, we know a wet regime will follow, yet we're also learning how to control the flow of water during drier times too. We know that fires will burn and we've manufactured firefighting marvels that can (eventually) stop giant walls of flames. But we have yet to own-up to the fact that the hand of humankind must take greater care in ensuring that our actions and accidents stop igniting the flames that need to be fought. And while we love the mountains that surround our homes, we too often forget that these giants may awaken, loosen their soil, rocks and trees, thus entombing all that we know, love, and care for.

We've come this far, and we can go much further. Placing proper attribution to the cause for disastrous effects focuses our attention toward the shortest possible paths for success. Distractions though from political leanings or uninformed climate cries set us backward in time. We can do better; we have done better; and we *must* continue to do better. Our world is changing; and so are we. Together, understanding the whole picture, we see not just the forest, but all of the intricate trees that comprise our wonderful, complex, intriguing, beautiful, and often confusing world.

Such is the story of human nature and our struggle to coexist with the elements of our world. It's an ancient story that has brought us from the trees and savannahs of Africa to become the ruling species of a planet. It's a story though, that tells how our dominion is just an illusion, hiding the fact that we are mere specks of biological matter in a universe much bigger and older than us, that we don't completely understand. It's a story of realizing that we must push forward if we want to progress; set aside our human arrogance; and learn from nature as well as our mistakes. It's a story that tells us we need to understand each other better, and work toward the same, common goals — for the wellbeing of our planet, and the sake of us all.

Such is the story of *Surf, Flood, Fire & Mud.*

Back Matter

Weather Charts

The next few pages show a few of the weather and wave charts from the more salient events talked about in this book. This isn't an exhaustive list of charts for each of the events in this story, but I've included what I felt stood out the most.

I opted to show the simple, basic, black-and-white weather charts from NOAA, to convey the contrast between what may seem like a benign drawing, and the magnitude of the event that it represents. There are no colors or hyped headlines, just annotations by the forecasters at NOAA. In the case of Montecito, I also included a satellite image, which, looking like just a simple stream of cotton, was in reality a terrifying weather system that ended up burying people alive.

Each of the following charts has an accompanying page to describe it, with a few notes for each one.

Sources for attaining these models, archived data, and storm databases then follows. The Notes section following that provides references to bibliographic information as well.

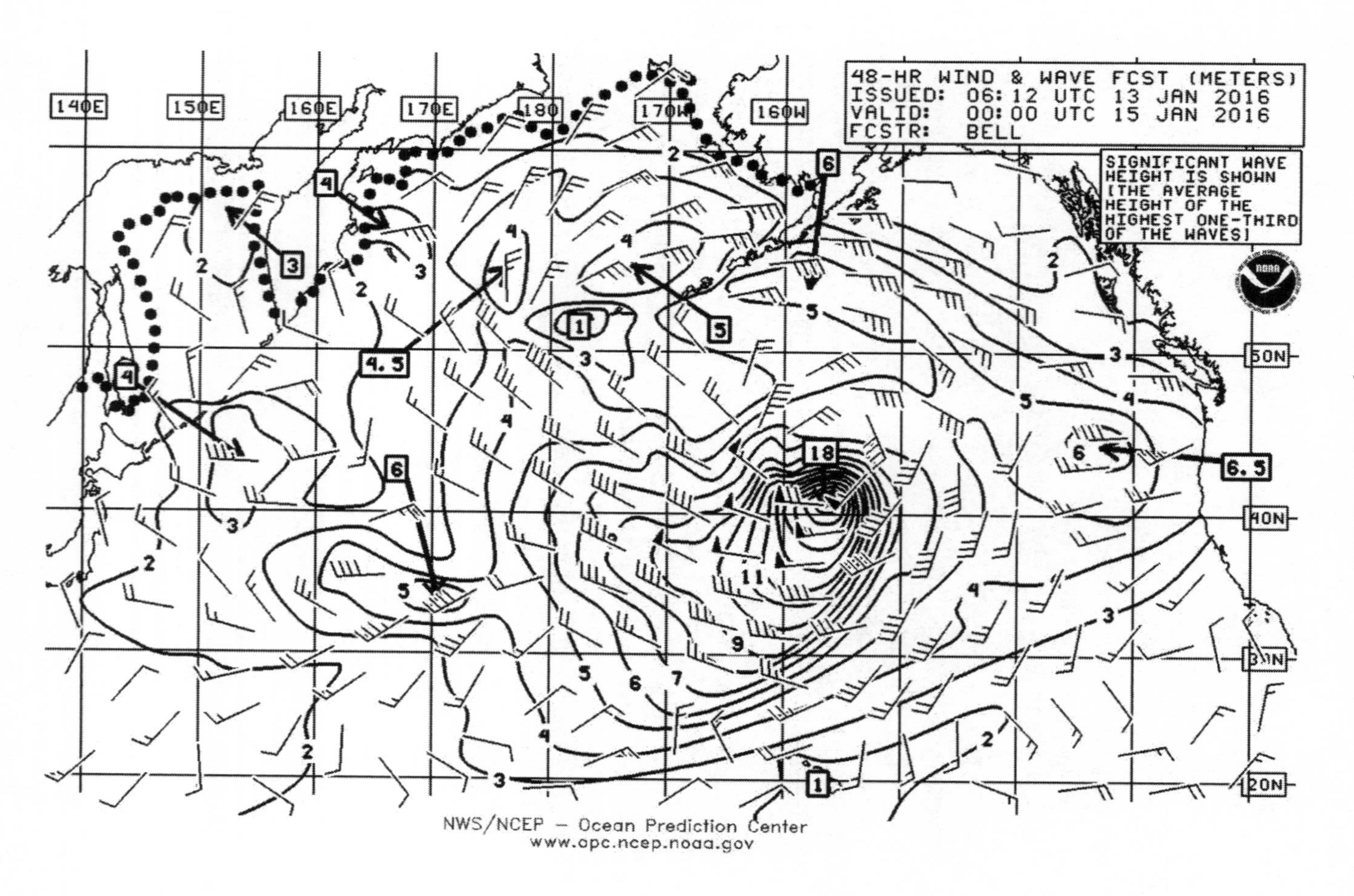
48-HR WIND & WAVE FCST (METERS)
ISSUED: 06:12 UTC 13 JAN 2016
VALID: 00:00 UTC 15 JAN 2016
FCSTR: BELL
SIGNIFICANT WAVE HEIGHT IS SHOWN (THE AVERAGE HEIGHT OF THE HIGHEST ONE-THIRD OF THE WAVES)
NWS/NCEP – Ocean Prediction Center
www.opc.ncep.noaa.gov
140E
150E
160E
170E
180
170W
160W
50N
40N
30N
20N

Monster Jaws

From "The Biggest Wave" section in the "Surf" chapter, this is the wave event that created the record breaking waves in early January 2016.

Near the center of this wave forecast chart you can see the strong fetch being created with 18 meter seas, which equals 59 feet. Hawaii is directly to the south, those little dots partially covered up by the square that has a "1" in it.

The wind barbs (what look like little flags) show that the strongest winds were blowing up to 70 kts (80 mph). While that may seem like anemic winds for such powerful seas — compared to a hurricane for instance — bear in mind that wave creation is a product of not just the strength of the wind, but how long a storm sits in one area. The longer a storm can sit in one place, more energy is transferred to the water, thus bigger waves are generated. This particular storm moved slower than most hurricanes, and over a much, much greater expanse of ocean as well. Having 80 mph winds while moving at a slow speed was enough to create some of the biggest seas I've seen in the North Pacific during my 20+ years as a forecaster.

Southern California is in the bottom right above the 30N mark. Note how the storm track moved this system north, due to high pressure near SoCal then. If not for that high pressure, this system would have given SoCal a much harder impact.

When Jaws got swell from this though, it was remarkable. Periods were 25 seconds, something almost unheard of from a northern hemisphere storm.

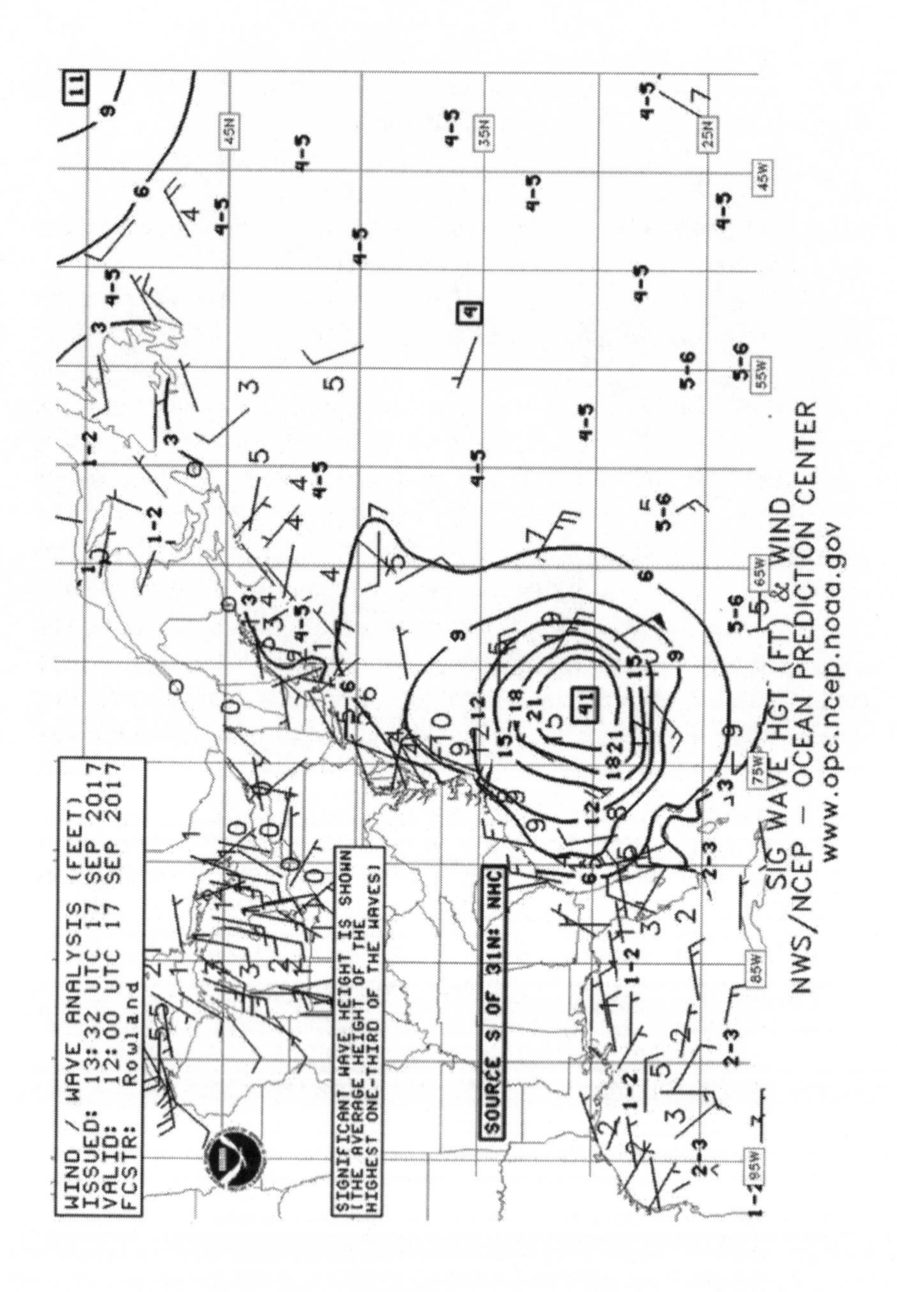
WIND / WAVE ANALYSIS (FEET)
ISSUED: 13:32 UTC 17 SEP 2017
VALID: 12:00 UTC 17 SEP 2017
FCSTR: Rowland
SIGNIFICANT WAVE HEIGHT IS SHOWN
(THE AVERAGE HEIGHT OF THE
HIGHEST ONE-THIRD OF THE WAVES)
SOURCE S OF 31N: NHC
SIG WAVE HGT (FT) & WIND
NWS/NCEP – OCEAN PREDICTION CENTER
www.opc.ncep.noaa.gov

East Coast Slammer

From "Between the Cracks" in the "Flood" chapter.

Hurricane Jose sits off the east coast of the United States with 41-foot seas at its center. Note that this wave chart shows feet, unlike the Pacific chart shown earlier for the Monster Jaws that is measured in meters.

Note the wind barbs (those flag looking things) show max winds reaching just 50 kts (58 mph). This differs slightly from the National Hurricane Center reports that showed, by this time, that winds were somewhere between 70-80 mph. Either way, those winds are much lower than the 155 mph winds that blew at Jose's peak. The 41-foot seas show once again that it's not just the strength of the wind, but how long it will blow over a given area that will determine how big the waves will be. Jose had originally been moving at 16 kts when it was blowing 120 mph winds, and about 12 kts when it was blowing the hellish, 150+ mph winds. But as those winds weakened to the ones shown in this chart, Jose was moving at a much slower 3 kts. Thus, the slower motion *and* winds made for powerful surf.

Although strong surf hit the east coast, Jose stayed out to sea. If it had made landfall it would have gotten much more attention.

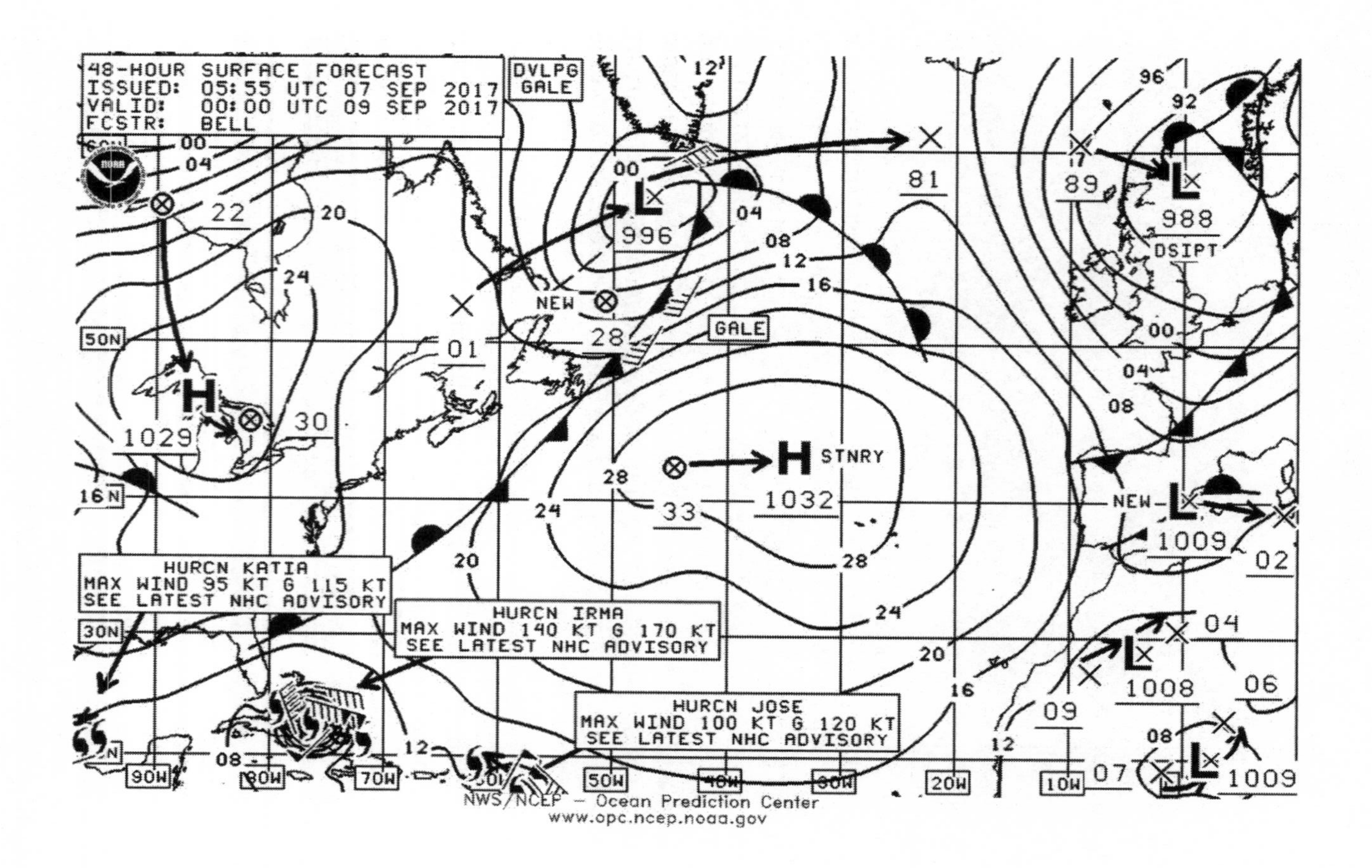
48-HOUR SURFACE FORECAST
ISSUED: 05:55 UTC 07 SEP 2017
VALID: 00:00 UTC 09 SEP 2017
FCSTR: BELL
DVLPG GALE
GALE
HURCN KATIA
MAX WIND 95 KT G 115 KT
SEE LATEST NHC ADVISORY
HURCN IRMA
MAX WIND 140 KT G 170 KT
SEE LATEST NHC ADVISORY
HURCN JOSE
MAX WIND 100 KT G 120 KT
SEE LATEST NHC ADVISORY
STNRY
1032
996
988
DSIPT
NEW
1029
1009
1008
NWS/NCEP – Ocean Prediction Center
www.opc.ncep.noaa.gov

Three Atlantic Hurricanes

From the "Flood" chapter, where three simultaneous hurricanes sit in hurricane alley: Katia, Irma, and Jose. This is something almost unheard of. During the El Niño in the summer of 2015, the Pacific had something similar, which would be more likely given the Pacific's much larger area. But for the Atlantic, this is eye-poppingly peculiar.

The wind barbs in this case are off the charts, showing 140 kt winds in Irma (two diamonds and 4 long lines on one of those wind barbs).

Also note that this Atlantic trio event occurred two years after El Niño when the Pacific was swinging toward a La Niña. As you may recall when talking about hurricanes earlier in the book, El Niño helps Pacific hurricanes but tends to kill Atlantic ones. The opposite is true for La Niña, which allows Atlantic hurricanes to form unabated. Thus, it is a contradiction to call for more frequent El Niños and more east coast-threatening storms under any climate changing conditions. However, more El Niños could mean more Pacific hurricanes, although those storms usually stay far out to sea, except in the case of Western Pacific typhoons.

Notice the big area of high pressure in the center of this chart. That's the Bermuda-Azores High mentioned at various times in the book, which helps to guide hurricanes from east to west across hurricane alley from its clockwise spin. It's also a key component to the North Atlantic Oscillation discussed in the "Niño y Nazaré" section of the "Surf" chapter.

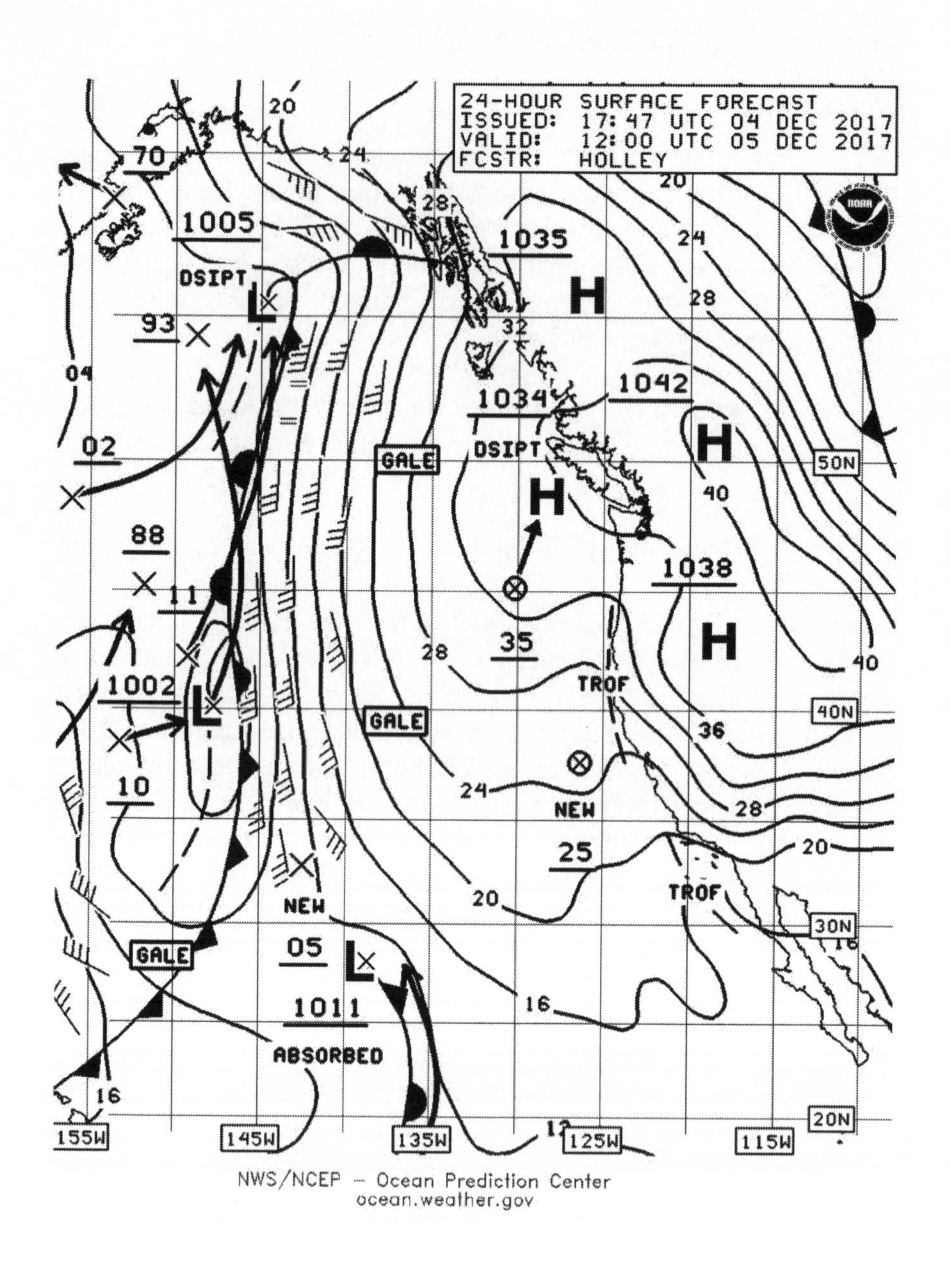
24-HOUR SURFACE FORECAST
ISSUED: 17:47 UTC 04 DEC 2017
VALID: 12:00 UTC 05 DEC 2017
FCSTR: HOLLEY
1005
OSIPT
1035
1034
1042
OSIPT
GALE
GALE
GALE
1038
35
TROF
1002
NEW
25
TROF
NEW
05
1011
ABSORBED
50N
40N
30N
20N
155W
145W
135W
125W
115W
NWS/NCEP – Ocean Prediction Center
ocean.weather.gov

The Wind Machine

From "The Thomas Fire" section in the "Fire" chapter. This is the wind event that spread the Thomas Fire.

In a classic Santa Ana pattern, high pressure dominates the American West, centered over the Great Basin. Lower pressure, in a trough, sits just off the coast. In this surface chart (measurements at sea level), the high pressure has a maximum center that measures 1042 millibars. The low pressure measures 1002 millibars. This is a great deal of difference, which creates a gradient between these two areas of pressure.

Note the strong gradients along the California coast as the high butts-up against a trough of lower pressure (the tight isobar lines over California). Notice also how tight those gradients are south of the high; this is where things get nasty. As the high spins clockwise and the low spins counterclockwise, a wind tunnel forms between them along those tight pressure gradients. On those southerly gradients, winds are blowing from east to west along those tight gradients/isobars, kicking up a strong Santa Ana over SoCal, centered over Ventura County.

One thing not shown in the chart is the low temperatures that were in the high, which produced temperature gradients as well, further fueling the winds.

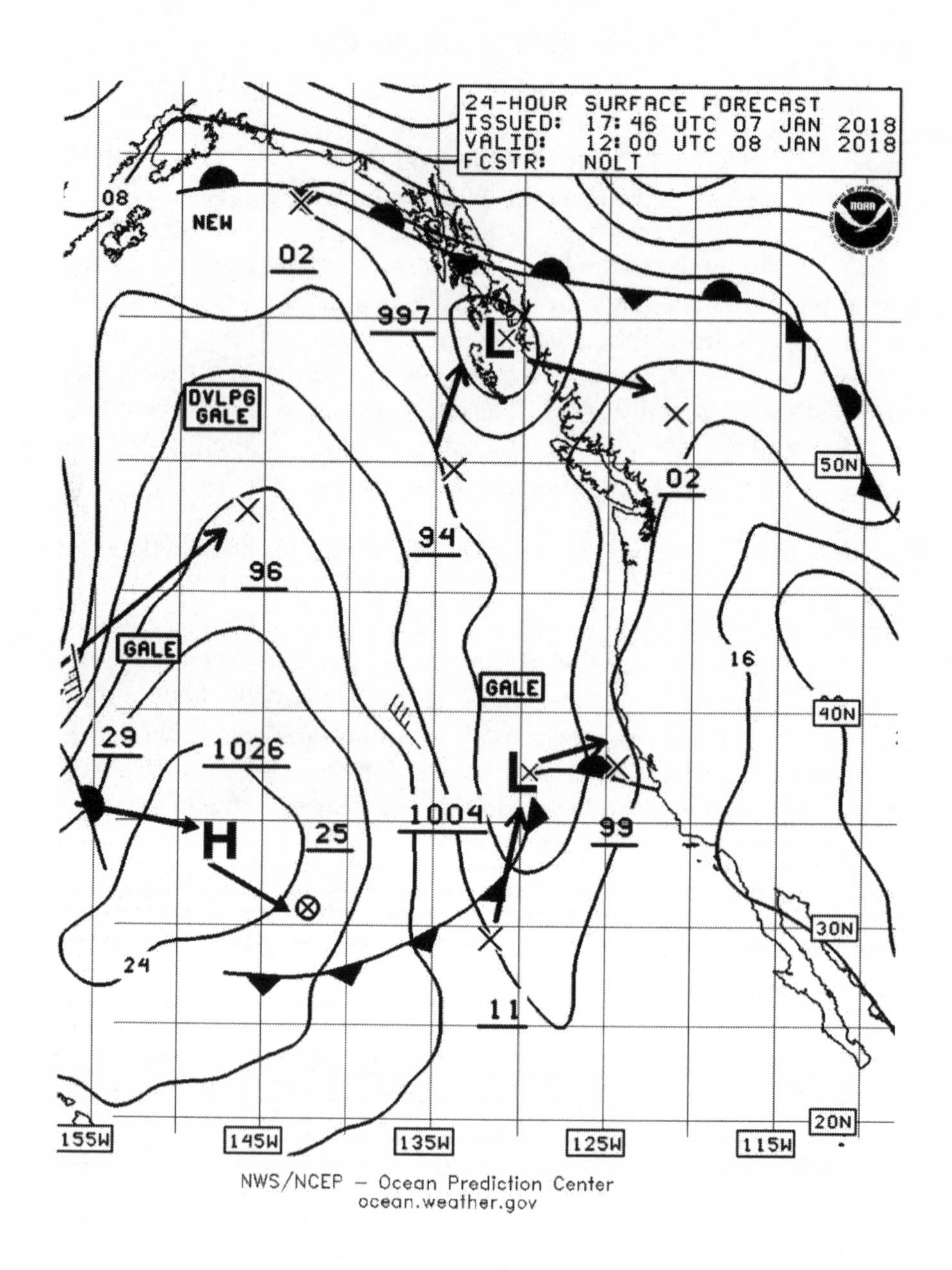
24-HOUR SURFACE FORECAST
ISSUED: 17:46 UTC 07 JAN 2018
VALID: 12:00 UTC 08 JAN 2018
FCSTR: NOLT
NEW
08
02
997
L
DVLPG GALE
94
96
02
50N
GALE
16
GALE
40N
29
1026
L
1004
H
25
99
30N
24
11
20N
155W
145W
135W
125W
115W
NWS/NCEP – Ocean Prediction Center
ocean.weather.gov

Montecito Slide

From "Montecito" in the "Mud" chapter. A thick plume of moisture made up the northeastern edge of a tropical moisture plume (image below). As a low pressure system approached (image left), it tapped into this plume, causing the copious rain that caused the mudslide.

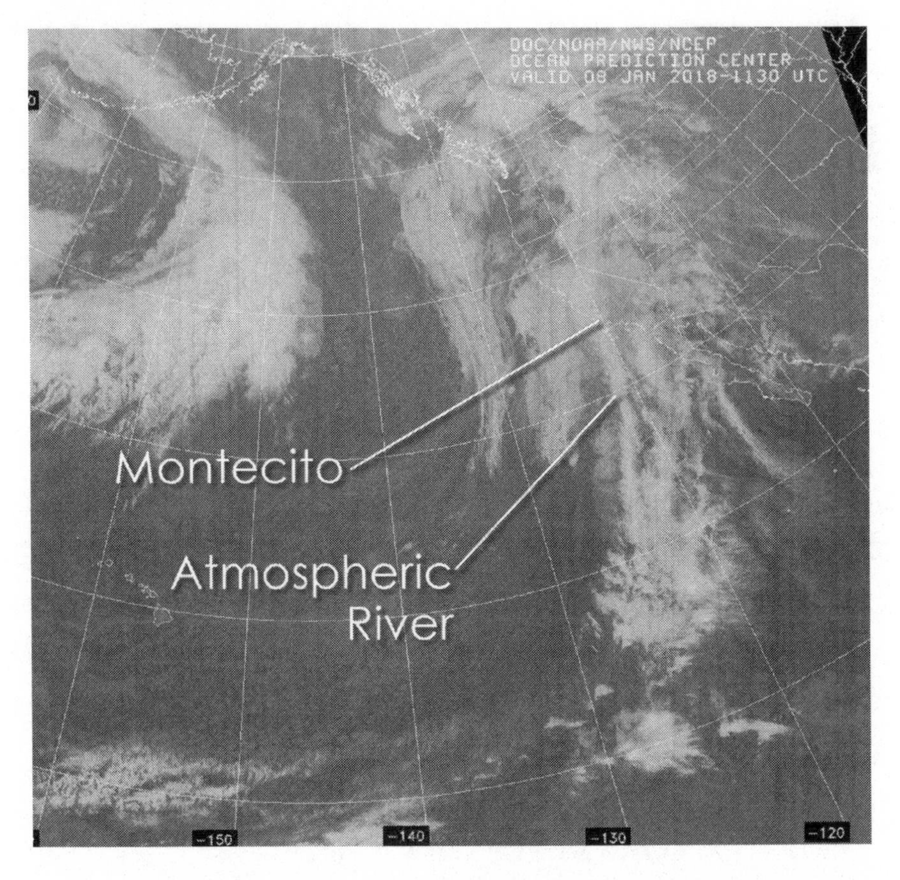

Notice how far south the low pressure trough extended (image left), well into Baja longitudes, which gave low latitude circulation into the plume.

Sources

I'm extremely grateful that NOAA and other agencies provide archived models and databases, which can be mined and analyzed to form the stories told here. The research for this book took countless hours looking over hundreds of weather models, charts, database collections, and historical accounts to provide accuracy for reporting these stories. Writing a non-fiction work can be tediously long, with sometimes hours of research and fact-checking just to get one sentence worded correctly. Having though the myriad sources available from NOAA and other agencies, the process of research was less painful than it would have been decades ago.

While there are footnote annotations throughout this book, I wanted to provide here links to publicly available information, which you can access free of charge to find weather models, data, and other such information.

An extremely useful starting point for finding archived weather models is through the National Centers for Environmental Information (NCEI), which you can access at:

https://www.ncei.noaa.gov/thredds/model/ncep_charts.html

...and also through:

https://www.ncdc.noaa.gov/data-access/model-data/model-datasets/service-records-retention-system-srrs

…or:

https://nomads.ncdc.noaa.gov/ncep-charts/new_charts/

You can also find information on extreme weather events through NOAA's NCEI at:

https://www.ncdc.noaa.gov/stormevents/

For wave models, CDIP has an archive at:

http://cdip.ucsd.edu/?nav=recent&sub=nowcast&xitem=get_model

Data sets can be retrieved through NCEI at:

https://www.ncdc.noaa.gov/cdo-web/search

Satellite and weather composites can be retrieved from SFSU at:

http://virga.sfsu.edu/crws/archive/sathts_pac_arch.html

Notes & Tidbits

---

[1] Surfer magazine “SWELLS THAT SHOOK THE WORLD”
Justin Housman Jan 9, 2014
https://www.surfer.com/features/winter-swells-history/

[2] Time magazine published June 24, 1974 stating that "…the atmosphere has been gradually cooler for the past three decades", and Newsweek April 28, 175 published an article titled "A Cooling World".

[3] While there were a number of scientists warning of the hype, Dr. Bjorn Lomborg fought to keep a level playing field with his *Cool It*, published in 2010, and points out arguments on a web site called "Bjorn Lomborg versus Al Gore at http://www.lomborg-errors.dk/Goreacknowledgederrors.htm

[4] A great tale of the polar bear's exploitation to be the ambassador can be found in *Wild Ones* by Jon Mooaliem. One of my favorite books of all time.

[5] South Park's season 19 uses a central character called PC Principal whose extreme PC views eventually lead to his downfall.

[6] Al Gore was quoted as saying "..I believe it is appropriate to have an over-representation of factual presentations on how dangerous it is as a predicate for opening up the audience to listen to what the solutions are.."
See https://en.wikiquote.org/wiki/Al_Gore#Quotes_about_Gore

[7] South Park loves to pick on Al Gore. If you haven't yet, I highly suggest watching the ManBearPig episode in the 10th season. It's funny as all get out, but also sad that Gore's exaggerations were too highly focused and shed a dark cloud on what he was trying to achieve.

[8] Technically speaking, this is known as “a neutral state of the ENSO”, which is a boring, drawn-out, tongue twister. To keep things simple, I just use the term “neutral Niño”.

[9] The Nobel Peace Prize 2007
https://www.nobelprize.org/nobel_prizes/peace/laureates/2007/

[10] ABC News “Al Gore's 'Inconvenient Truth'? -- A $30,000 Utility Bill” Jake Tapper Feb 26, 2007
http://abcnews.go.com/Politics/GlobalWarming/story?id=2906888

[11] Wikipedia, El Niño,
https://en.wikipedia.org/wiki/El_Ni%C3%B1o#Cultural_history_and_prehistoric_information

[12] Wikipedia Peruvian anchoveta, 2/2016

[13] NOAA NWS, CPC, Cold & Warm Episodes by Season
Retrieved 2/26/2006 from:

---

http://www.cpc.ncep.noaa.gov/products/analysis_monitoring/ensostuff/ensoyears.shtml

[14] NOAA, Billy Kessler, Pacific Marine Environment Laboratory, FAQ 3
Retrieved 2/25/2006 from:
http://www.pmel.noaa.gov/%7Ekessler/occasionally-asked-questions.html

[15] Geophysical Research Letters, AGU Journal, "Causes and impacts of the 2014 warm anomaly in the NE Pacific"
May, 5, 2015, Bond et al
http://onlinelibrary.wiley.com/doi/10.1002/2015GL063306/full

[16] Wikipedia "Ridiculously Resilient Ridge"
https://en.wikipedia.org/wiki/Ridiculously_Resilient_Ridge

[17] Wikipedia "The Blob", Cause
https://en.wikipedia.org/wiki/The_Blob_(Pacific_Ocean)#Cause

[18] Model archive records for pressure and sea analysis from NOAA NDCD at:
https://www.ncdc.noaa.gov/data-access/model-data/model-datasets/service-records-retention-system-srrs

[19] AccuWeather Dec 13, 2015 "High Surf, Coastal Flooding Unundates California Coast"
https://www.accuweather.com/en/weather-news/high-surf-coastal-flooding-inu/54135501

[20] ABC7 Dec 11, 2015 Strong surf pounds Sounland beaches, causes Ventura pier closure
http://abc7.com/weather/ventura-pier-closes-due-to-damage-from-high-surf/1118907/

[21] NCDC Storm Database San Diego High Surf Alert December 11, 2015
https://www.ncdc.noaa.gov/stormevents/eventdetails.jsp?id=608516

[22] FOX5 "King tide floods La Jolla Shores"
http://fox5sandiego.com/2015/12/23/high-tide-floods-la-jolla-shores/

[23] While I realize a fetch with 59' seas is hard to comprehend, I thought it worth a note here on how that was derived. It is from the 48-hour Wind & Wave Forecast by NOAA that was issued on Jan 13, 2016 that shows 18 meter wave heights. I provide links to model archives where you can find charts like this in the Acknowledgments chapter.

[24] CDIP: Waimea Bay Buoy (106) 2016-01-15 10.33 meters, or 34 feet.

[25] Guinness World Records, Largest wave surfed (paddle-in)
http://www.guinnessworldrecords.com/world-records/78561-largest-wave-surfed-paddle-in

[26] See, I told you it was big! Those 59' seas turned into 80' breaking waves. Source: ESPN, "Swallowed by Jaws" Jan 26, 2016 Lucas Gilman
http://www.espn.com/espn/feature/story/_/id/14626909/surfers-risk-death-paddling-historic-80-foot-waves-jaws

[27] Guinness World Records, "Monday Motivation: Garrett McNamara, King of the Surf"

Kristen Stephenson Mar 20, 2017
http://www.guinnessworldrecords.com/news/2017/3/monday-motivation-garrett-mcnamara-king-of-the-surf

[28] There is some confusion with whether Guinness considers McNamara's Nazaré 200-footer to be official. It was rode three years prior to Gold's paddle-in wave at Jaws, but Guinness doesn't seem to have confirmed that, even though they had confirmed Gold's more recent ride. See:
guinnessworldrecords.com/world-records/78115-largest-wave-surfed-unlimited
and
guinnessworldrecords.com/world-records/78561-largest-wave-surfed-paddle-in
and
guinnessworldrecords.com/news/2017/3/monday-motivation-garrett-mcnamara-king-of-the-surf

[29] "The Nazaré Coast, the Submarine Canyon and the Giant Waves" Universidade De Coimbra, February 2015, Pedro Cunha and Margarida Porto Gouveia
https://www.researchgate.net/publication/275522569_The_Nazare_coast_the_submarine_canyon_and_the_giant_waves_-_a_synthesis

[30] NOAA Climate Prediction Center, Teleconnections, North Atlantic Oscillation
http://www.cpc.ncep.noaa.gov/products/precip/CWlink/pna/month_nao_index.shtml

[31] NOAA/NESDID SST Anomaly 2/15/2016
http://www.ospo.noaa.gov/data/sst/anomaly/2016/anomnight.2.15.2016.gif

[32] CBS News 5 KPIX "Santa Cruz Native Nic Lamb Wins 2016 Titans Of Mavericks Surf Contest" Feb 12, 2016
http://sanfrancisco.cbslocal.com/2016/02/12/santa-cruz-native-nic-lamb-wins-2016-titans-of-mavericks-surf-contest/

[33] Surfer Today, "The most notable deaths in surfing" retrieved 2/10/18 from:
https://www.surfertoday.com/surfing/13640-the-most-notable-deaths-in-surfing

[34] The Guardian "Shark nets to be trialled again on New South Wales north coast beaches" Christopher Knaus Aug 31, 2017
https://www.theguardian.com/australia-news/2017/sep/01/shark-nets-to-be-trialled-again-on-new-south-wales-north-coast-beaches

[35] Florida Museum of Natural History, University of Florida, International Shark Attack File (ISAF)
Accessible at http://www.flmnh.ufl.edu/fish/isaf/shark-attacks-maps-data/north-america/california

[36] ABC News, September 6, 2015 "Kayaker Bitten by Shark off Malibu"
http://abcnews.go.com/US/kayaker-bitten-shark-off-malibu-beach-respect/story?id=33563768

[37] KTLA Channel 5 News, Los Angeles, September 17, 2015, "Video Captures Aggressive Encounter with Hammerhead Shark off Dana Point"

http://ktla.com/2015/09/17/video-captures-aggressive-encounter-with-hammerhead-shark-off-dana-point/

[38] I actually took up hiking for quite some time to chronical the development of El Niño and the drought shooting high-res panoramas. If you get bored with this book or want a break, you can see my handiwork at www.NathanCoolPhoto.com. In any event, I felt safer climbing local mountains than wondering what hungry creature may be stirring some distance below me.

[39] Orange County Register, "15 sharks seen off O.C." May 10, 2017 https://www.ocregister.com/2017/05/10/15-sharks-spotted-by-oc-sheriffs-helicopter-near-beachgoers-prompts-warning-from-the-sky-exit-the-water-in-a-calm-manner/

[40] KTLA Channel 5 News, Los Angeles, October 15, 2015, "Incredibly Venomous' Yellow-Bellied Sea Snake Seen in California for 1st Time in 30 Years"
Accessible at http://ktla.com/2015/10/16/incredibly-venomous-yellow-bellied-sea-snake-seen-in-california-for-1st-time-in-30-years/

[41] National Geographic, "What's This Tropical, Venomous Sea Snake Doing in California?" Brian Clark Howard Dec 21, 2015
https://news.nationalgeographic.com/2015/12/151221-yellow-bellied-sea-snake-california/

[42] The San Diego Tribune, April 15, 2014 "El Nino could mean epic sport fishing" http://www.sandiegouniontribune.com/news/2014/apr/15/outdoors-el-nino-fishing-san-diego/

[43] Observations of Fishes Associated with the 1997-98 El Niño off California, by the California Dept. of Fish and Game, Robert N. Lea and Richard H. Rosenblatt, Vol. 41, 2000
Accessible at
http://web.calcofi.org/publications/calcofireports/v41/Vol_41_Lea___Rosenblatt.pdf

[44] CBS News, June 17, 2015 "Millions of red tuna crabs invade California beaches" http://www.cbsnews.com/news/millions-red-tuna-crabs-invade-california-beaches/

[45] CBS KPIX, June 12, 2015 "Return of the Blob! Big Purple Sea Slugs Wash Ashore in East Bay"
http://sanfrancisco.cbslocal.com/2015/06/12/big-purple-sea-slugs-wash-ashore-in-east-bay/

[46] WWF, "Crisis in global oceans as populations of marine species halve in size since 1970" September 16, 2015
http://www.wwf.org.uk/about_wwf/press_centre/?unewsid=7673

[47] The stories of Easter Island and Greenland Norse collapse is detailed well in Jared Diamond's "Collapse: why societies choose to fail or succeed". A highly recommended read.

[48] The journal *Science*, January 22, 2016 p.321, "No more whaling reviews"

---

[49] The journal *Science*, January 22, 2016 p.323, "Shell trade pushes giant clams to the brink"

[50] The Guardian, "Thousands of squid washed ashore in Chile raise health concerns" January 14, 2016
http://www.theguardian.com/environment/video/2016/jan/15/thousands-of-squid-washed-ashore-in-chile-raise-health-concerns-video

[51] U.S. News & World Report, "Alaska Seabirds Are Likely Starving to Death" January 12, 2016
http://www.usnews.com/news/science/news/articles/2016-01-12/starvation-suspected-in-massive-die-off-of-alaska-seabirds

[52] LA Times, "Guadalupe fur seals dying at an alarming rate off California coast" October 1, 2015
http://www.latimes.com/local/lanow/la-me-ln-guadalupe-fur-deals-dying-20151001-story.html

[53] The El Niño of 1991-92 peaked in January 1992 with a peak SST anomaly of 1.7°C, about 35% lower than the 2015-16 El Niño, and 30% lower than the 1997-98 El Niño.

[54] NASA has a really cool satellite shot of these three storm at:
https://earthobservatory.nasa.gov/IOTD/view.php?id=86512

[55] Climate Central, "Hawaii May Say 'Aloha' to More Hurricanes"
Andrea Thompson Aug 5, 2015
http://www.climatecentral.org/news/warming-hawaii-hurricanes-guillermo-19313

[56] NOAA, Geophysical Fluid Dynamics Laboratory, "Global Warming and Hurricanes: An Overview of Current Research Results" Jan 24, 2018
https://www.gfdl.noaa.gov/global-warming-and-hurricanes/

[57] IPCC AR5 "The Physical Science Basis" Chapter 10, page 871.

[58] Los Angeles Times, December 9, 2015 Massive El Niño is now 'too big too fail'
http://www.latimes.com/local/lanow/la-me-ln-massive-el-nino-is-now-too-big-to-fail-scientist-says-20151009-story.html

[59] ABC7 October 13, 2015 Monstrous El Niño expected to hit Southern California
http://abc7.com/weather/monstrous-el-nino-expected-to-hit-southern-california/1032185/

[60] Los Angeles Times, December 9, 2015 Massive El Niño is now 'too big too fail'
http://www.latimes.com/local/lanow/la-me-ln-massive-el-nino-is-now-too-big-to-fail-scientist-says-20151009-story.html

[61] The Weather Channel, NASA Scientist to California: 'You'll Essentially Be Riding in Your Kayak' When El Niño Sets In
https://weather.com/science/environment/news/bill-patzert-el-nino

[62] Drought Monitor, University of Nebraska-Lincoln
http://droughtmonitor.unl.edu

[63] U.S. News "California's Unprecedented Water Cuts: A Primer" Alan Neuhauser April 2, 2015
https://www.usnews.com/news/articles/2015/04/02/global-warming-spurs-california-water-restrictions

[64] KCET, "Drought by the Numbers: Where Does California Water Go?" D.J. Waldie Feb 10, 2014
https://www.kcet.org/commentary/drought-by-the-numbers-where-does-california-water-go

[65] NOAA California Nevada River Forecast Center, Heavy Precipitation Event Southern California February 17-12, 2015. Here NOAA details some of the record rain events in SoCal.
http://www.cnrfc.noaa.gov/storm_summaries/feb2005storms.php

[66] Royal Meteorological Society, "The global climate anomaly 1940–1942" Stefan Brönnimann Institute for Atmospheric and Climate Sciences, ETH, Zürich, Switzerland December 2005
http://onlinelibrary.wiley.com/doi/10.1256/wea.248.04/pdf

[67] The Telegraph, "Second World War: Frozen to death by the Fuhrer" Andrew Roberts, July 25, 2009
http://www.telegraph.co.uk/history/britain-at-war/5907564/Second-World-War-Frozen-to-death-by-the-Fuhrer.html

[68] Causes and impacts of the 2014 warm anomaly in the NE Pacific
Geophysical Research Letters, May 5, 2015 Bond et. al.
http://onlinelibrary.wiley.com/doi/10.1002/2015GL063306/full

[69] Pasadena Star "A retiring Bill Patzert, JPL's 'Prophet of California Climate,' leaves behind a legacy of ocean research and media appearances" Steve Scauzilo Jan 14, 2018
https://www.pasadenastarnews.com/2018/01/14/a-retiring-bill-patzert-jpls-prophet-of-california-climate-leaves-behind-a-legacy-of-ocean-research-and-media-appearances/

[70] Great Flood of 1862, WikiPedia
https://en.wikipedia.org/wiki/Great_Flood_of_1862

[71] Climate Since AD 1500, Quinn and Neal Historical record of El Niño events
NOAA Paleoclimatology Program
ftp://ftp.ncdc.noaa.gov/pub/data/paleo/climate1500ad/ch32.txt

[72] Wikipedia "Christmas flood of 1964"
https://en.wikipedia.org/wiki/Christmas_flood_of_1964

[73] University of Arkansas, Professor David Stahle
Tree-Ring Data Show History, Pattern to Droughts, Feb 15, 2013
https://news.uark.edu/articles/20244/tree-ring-data-show-history-pattern-to-droughts

and
Causes and consequences of nineteenth century droughts in North America
Richard Seager and Cline Herweijer, Lamont-Doherty Earth Observatory of Columbia University
http://ocp.ldeo.columbia.edu/res/div/ocp/drought/nineteenth.shtml

[74] Wikipedia "1964 Pacific typhoon season"
https://en.wikipedia.org/wiki/1964_Pacific_typhoon_season

[75] American Meteorological Society, Department of Atmospheric Science, Colorado State University, Fort Collins, "All-Season Climatology and Variability of Atmospheric River Frequencies over the North Pacific", Mundhenk et. al. June 16, 2016
http://journals.ametsoc.org/doi/full/10.1175/JCLI-D-15-0655.1

[76] For the AR event of 1993: American Meteorological Society Journals Online, "Atmospheric Rivers and Cool Season Extreme Precipitation Events in the Verde Basin of Arizona", Table 2, by Erick Rivera, Francina Dominguez, Christopher Castro
https://journals.ametsoc.org/doi/pdf/10.1175/JHM-D-12-0189.1

[77] Lawrence Livermore National Laboratory, "Increase in atmospheric moisture tied to human activities", Anne M. Stark, Sep. 18, 2007
https://www.llnl.gov/news/increase-atmospheric-moisture-tied-human-activities

[78] Science Magazine (AAAS), California rains put spotlight on atmospheric rivers, Julia Rosen Feb. 22, 2017
http://www.sciencemag.org/news/2017/02/california-rains-put-spotlight-atmospheric-rivers

[79] Scripps Institution of Oceanography at UC San Diego
Center for Western Weather and Water Extremes
"How Many Atmospheric Rivers Have Hit the U.S. West Coast During the Remarkable Wet Water Year 2017?"
http://cw3e.ucsd.edu/how-many-atmospheric-rivers-have-hit-the-u-s-west-coast-during-the-remarkably-wet-water-year-2017/

[80] NOAA Madden-Julian Oscillation: Recent Evolution, Current Status and Predictions, July 13, 2015
http://www.cpc.ncep.noaa.gov/products/precip/CWlink/MJO/ARCHIVE/PDF/mjo_evol-status-fcsts-20150713.pdf

[81] The Journal Nature, "Extratropical Forcing Triggered the 2015 Madden–Julian Oscillation–El Niño Event" Chi-Cherng Hong, Huang-Hsiung Hsu, Wan-Ling Tseng, Ming-Ying Lee, Chun-Hoe Chow & Li-Chiang Jiang
April 24, 2017
https://www.nature.com/articles/srep46692

[82] The American Meteorological Society, "Does the Madden–Julian Oscillation Influence Wintertime Atmospheric Rivers and Snowpack in the Sierra Nevada?"
Bin Guan and Duane E. Waliser, Jet Propulsion Laboratory, California Institute of Technology, Pasadena, California

Noah P. Molotch, Department of Geography and Institute for Arctic and Alpine Research, University of Colorado, Boulder, Colorado, and Jet Propulsion Laboratory, California Institute of Technology, Pasadena, California
Eric J. Fetzer, Jet Propulsion Laboratory, California Institute of Technology, Pasadena, California
Paul J. Neiman, Physical Sciences Division, NOAA/Earth System Research Laboratory, Boulder, Colorado
Feb 1, 2012
https://journals.ametsoc.org/doi/abs/10.1175/MWR-D-11-00087.1

[83] NOAA Madden-Julian Oscillation: Recent Evolution, Current Status and Predictions, Feb 6, 2017 and Feb 20, 2017
http://www.cpc.ncep.noaa.gov/products/precip/CWlink/MJO/ARCHIVE/PDF/mjo_evol-status-fcsts-20170206.pdf
and
http://www.cpc.ncep.noaa.gov/products/precip/CWlink/MJO/ARCHIVE/PDF/mjo_evol-status-fcsts-20170220.pdf

[84] NOAA National Centers for Environmental Information
https://www.ncdc.noaa.gov/stormevents/eventdetails.jsp?id=676363

[85] California Data Exchange Center, Dept. of Water Resources
https://cdec.water.ca.gov/queryMonthly.html

[86] NASA Earth Observatory, Water Levels Rise on Shasta Lake
https://earthobservatory.nasa.gov/NaturalHazards/view.php?id=87822

[87] On April 7, 2017, the Governor's Office issued a press release where Governor Brown said "This drought emergency is over, but the next drought could be around the corner. Conservation must remain a way of life." He was right. There doesn't seem to be though any news coverage on this, except for brief references; after all, media will "tell you 'bout the plane crash with a gleam in her eye…give us dirty laundry." (Google that with "Don Henley" if you are not a music fan from the 1980s). Brown's press release can be found at:
https://www.gov.ca.gov/2017/04/07/news19747/

[88] Wikipedia, Oroville Crisis
https://en.wikipedia.org/wiki/Oroville_Dam_crisis

[89] KQED News, "Feds Ask State to Explain Cracks in New Oroville Spillway Concrete" Dan Brekke Nov 27, 2017
https://ww2.kqed.org/news/2017/11/27/feds-ask-state-to-explain-cracks-in-new-concrete-on-oroville-spillway/

[90] NewsDeeply, "Oroville Dam Incident Explained: What Happened, Why and What's Next" Ian Evans June 14, 2017 at:
https://www.newsdeeply.com/water/articles/2017/06/14/oroville-dam-incident-explained-what-happened-why-and-whats-next

[91] Wikipedia "2017 California floods"

https://en.wikipedia.org/wiki/2017_California_floods

[92] NOAA National Centers for Environmental Information, Storm Events Database https://www.ncdc.noaa.gov/stormevents/eventdetails.jsp?id=680852

[93] Oroville Dam Spillway Incident Independent Forensic Team report Jan 5, 2018 https://damsafety.org/sites/default/files/files/Independent%20Forensic%20Team%20Report%20Final%2001-05-18.pdf

[94] NOAA, Climate.gov "Maya Express behind Gulf Coast soaking", Tom Di Liberto March 22, 2016
https://www.climate.gov/news-features/event-tracker/maya-express-behind-gulf-coast-soaking

[95] Houston Chronicle, "Gore: Global warming will cause more hurricanes like Harvey" Alex Stuckey Oct 23, 2017
https://www.chron.com/news/article/Al-Gore-global-warming-will-cause-more-Hurricane-12300688.php

[96] IPCC TAR, Summary for Policy Makers, page 14, "Many models project more El Niño-like mean conditions in the tropical Pacific." and page 6 table SPM-1, "El Niño Events"
https://www.ipcc.ch/pdf/climate-changes-2001/synthesis-syr/english/summary-policymakers.pdf
Also, IPCC AR5 Summary for Policy Makers, page 23, "Due to the increase in moisture availability, ENSO related precipitation variability on regional scales will likely intensify. Natural variations of the amplitude and spatial pattern of ENSO are large and thus confidence in any specific projected change in ENSO and related regional phenomena for the 21st century remains low."
https://www.ipcc.ch/pdf/assessment-report/ar5/wg1/WG1AR5_SPM_FINAL.pdf

[97] IPCC TAR, Summary for Policy Makers, page 15 table SPM-2 "Increase in tropical cyclone peak wind intensities, mean and peak precipitation intensities"
https://www.ipcc.ch/pdf/climate-changes-2001/synthesis-syr/english/summary-policymakers.pdf
and IPCC AR5 page 1220 "Cyclones"
https://www.ipcc.ch/pdf/assessment-report/ar5/wg1/WG1AR5_Chapter14_FINAL.pdf

[98] National Hurricane Center Harvey discussion 10:00 AM CDT Sat Aug 26, 2017

[99] The Texas Tribune "Can flooded-out Houstonians win lawsuits against Army Corps?" Sep 28, 2017 Kiah Collier
https://www.texastribune.org/2017/09/28/will-flooded-out-houstonians-prevail-lawsuits-against-army-corps/
Also: KCRA "Levee that had been breached is fortified?, say officials" Aug 29, 2017 AP
http://www.kcra.com/article/residents-south-of-houston-urged-to-leave-area-after-levee-breach/12118734
And Wikipedia https://en.wikipedia.org/wiki/Hurricane_Harvey

[100] Wikipedia, "List of wettest tropical cyclones in the U.S." 1/23/18

https://en.wikipedia.org/wiki/List_of_wettest_tropical_cyclones_in_the_United_States

[101] CNN "Houston knew it was at disk of flooding, so why didn't the city evacuate?" Dakin And one August 29, 2017
https://www.cnn.com/2017/08/27/us/houston-evacuation-hurricane-harvey/index.html

[102] Wikipedia "Hurricane Irma". Path and strength referenced through NHC archives.

[103] NOAA, Geophysical Fluid Dynamics Laboratory, "Global Warming and Hurricanes: An Overview of Current Research Results" Jan 24, 2018
https://www.gfdl.noaa.gov/global-warming-and-hurricanes/

[104] The New York Times "Puerto Rico Declares a Form of Bankruptcy" Mary Williams Walsh, May 3, 2017
https://www.nytimes.com/2017/05/03/business/dealbook/puerto-rico-debt.html

[105] New York Post "'Inept' Puerto Rican government 'riddled with corruption': CEO" Jorge Rodriquez Sep 30, 2017
https://nypost.com/2017/09/30/inept-puerto-rican-government-riddled-with-corruption-ceo/

[106] CNN, "About 1 million Americans without running water. 3 million without power. This is life one month after Hurricane Maria" John D. Sutter Oct 20, 2017
http://edition.cnn.com/2017/10/18/health/puerto-rico-one-month-without-water/index.html

[107] The New York Times "What Puerto Rico Is, and Isn't, Getting in Disaster Relief" PATRICIA MAZZEI FEB. 8, 2018
https://www.nytimes.com/2018/02/08/us/puerto-rico-disaster-relief.html

[108] Google Trends comparing Jose, Maria, and Irma interest from 9/1/17 to 9/20/17 retrieved 2/9/18 at
https://trends.google.com/trends/explore?date=2017-09-01%202017-09-30&q=hurricane%20jose,hurricane%20maria,hurricane%20irma

[109] Harvey cost $125B and had 69 deaths; Irma cost $64B and had 66 deaths; and Maria cost $92 and had 112 deaths. Taken from Wikipedia pages for each storm retrieved 2/9/2018.

[110] You can see this picture at NASA's website at:
https://earthobservatory.nasa.gov/NaturalHazards/view.php?id=90931

[111] Irma cost $64,000,000,00 and Jose caused $2,800,000 in damages, actually 0.4% but rounded up.

[112] NOAA Climate.gov, "Very wet 2017 water year ends in California" October 1, 2017 Tom Di Liberto
https://www.climate.gov/news-features/featured-images/very-wet-2017-water-year-ends-california

---

[113] USDA Forest Service, "Fire Ecology of Ponderosa Pine and the Rebuilding of Fire-Resilient Pondersa Pine Ecosystems" Stephen A. Fitzgerald, 2015 https://www.fs.fed.us/psw/publications/documents/psw_gtr198/psw_gtr198_n.pdf

[114] Ventura County Fire Department, Fire Hazard Reduction Program http://vcfd.org/images/prevention/Brochures/VCFD_Fire_Hazard_Reduction_Brochure_final_2017.pdf

[115] KPCC "'Defensible space' couldn't keep Thomas Fire from burning Ventura County" Emily Guerin Dec 19, 2017. Interviewing veteran firefighter Steve Kaufmann, he stated "Defensible space has a very specific use. It's to provide a place for firefighters to do their work. It doesn't actually necessarily in and of itself protect the home from ignition" Source:
https://www.scpr.org/news/2017/12/19/79035/defensible-space-couldn-t-keep-thomas-fire-from-bu/

[116] Climate change has an additional, highly likely relation to fires and hurricanes as well. For this discussion though, there is a definitive link of El Niño to these events. If we go further down the rabbit hole, climate change can be seen, but is not as relevant here for realize the chain reaction of events.

[117] NOAA National Centers for Environmental Information, "Billion-Dollar Weather and Climate Disasters"
https://www.ncdc.noaa.gov/billions/events/US/1980-2017

[118] The Thomas Fire cost of about $300 million

[119] Journal Science, Policy Forum "Reform forest fire management" Sep 18, 2015, North et.al.
http://science.sciencemag.org/content/349/6254/1280

[120] Journal Science, "Who is starting all those wildfires? We are" Warren Cornwall Sep 12, 2017
http://www.sciencemag.org/news/2017/09/who-starting-all-those-wildfires-we-are

[121] U.S. National Park Service, Fire and Aviation Management, "Large Fires & Fatalities" https://www.nps.gov/fire/wildland-fire/learning-center/fireside-chats/history-timeline.cfm

[122] Population.us, Jan 26, 2018
http://population.us/ca/san-buenaventura-ventura/

[123] Lamont-Doherty Earth Observatory, The Earth Institute at Columbia University, Richard Seager and Celine Herweijer
http://ocp.ldeo.columbia.edu/res/div/ocp/drought/nineteenth.shtml

[124] Wikipedia, Rocky Montain locust
https://en.wikipedia.org/wiki/Rocky_Mountain_locust

[125] History Net, "1874: The Year of the Locust"
http://www.historynet.com/1874-the-year-of-the-locust.htm

[126] Los Angeles Times archive, "Fire the price we pay" October 15, 2008 Tim Rutten http://articles.latimes.com/2008/oct/15/opinion/oe-rutten15

[127] "Santa Ana" winds are generally thought to be a mispronunciation of the Spanish word "Santana", Satan in English. That is disputed from an ad that appeared in the LA Times in 1881 where a land seller said this plot had "no fogs or Santa Ana winds". That though was also disputed by readers at that time who insisted that the advertiser miswrote the true spelling. The sticking point for this dispute is that, even from the late 1800s, many argued that the name is "Santa Ana" due to the winds passing through the Santa Ana canyon in Orange County, where the 91 freeway runs. That though is questionable since that strongest and widest spread offshore winds blow between the San Gabriels and the San Emigdio mountains through Ventura County, and in Orange County those winds are channeled between the San Gabriels and the San Bernardino mountains. That Santa Ana Canyon, while known for its offshore winds, is a smaller wind funneled area than the other mountain passes. Additionally, Northern Californians call their winds Diablo winds, for Devil, which too is questioned whether it was named from a canyon or the devil. While it's quite likely that the SoCal winds were originally called Santana, no one will probably ever know the true pronunciation; however, if you've ever experienced them, I'm sure you'd consider them to be blown from the breath of a demon.

[128] University of Utah, Department of Atmospheric Sciences, Meso West archives http://mesowest.utah.edu/

[129] Climate Dynamics, "El Niño-Southern Oscillation impacts on winter winds over Southern California", Berg et. al. July 26, 2012, Department of Atmospheric and Oceanic Sciences, UCLA and Earth System Research Laboratory, NOAA https://link.springer.com/article/10.1007%2Fs00382-012-1461-6

[130] AGU Publications, "Santa Ana Winds of Southern California: Their climatology, extremes, and behavior spanning six and a half decades". Guszan-Morales et. al. Jan 27, 2016 Scripps Institution of Oceanography UCSD, Theiss Rsearch La Jolla, Cooperative Institute for Research in Environmental Science Boulder, NOAA, and USGS https://scripps.ucsd.edu/programs/cnap/wp-content/uploads/sites/109/2017/02/GuzmanMorales2016_SantaAnaWinds.pdf

[131] IPCC Fifth Assessment Report (AR5) , Summary for Policymakers, Climate Change 2014, Impacts, Adaptation, and Vulnerability.

[132] IPCC AR5, Working Group 2, Impacts, Adaptations and Vulnerability, sections 15.2.2.1.2 through 15.2.2.1.5, but also throughout chapter 15, North America.

[133] Even in the IPCC's AR4, when talking about Wind Speed (section 11.3.3.5), there is no mention of North America at all.

[134] IPCC AR5, The Physical Science Basis, Chapter 10, page 871, "Climate Extremes"

[135] Los Angeles Daily News "Could 'firefighting at Christmas' become the new normal in California?" Elliot Spagat and Brian Melley Dec 9, 2017 https://www.dailynews.com/2017/12/09/could-firefighting-at-christmas-become-the-new-

normal-in-california/

[136] Climate change is often linked to droughts and severe rain. SoCal got both, but the areas that are to be affected, according to the IPCC, are vastly different…you don't get both in the same area. Additionally, the Great Flood of 1861-62 was well before climate change was underway as well, so one could argue that the heavy rains of 2016-17 were the same deal, and not climate change related. The fact is, climate change is linked to many things, but you have to be careful what you attribute it to, since, sometimes, stuff just happens. If you blame everything on climate change, then it becomes less believable. Climate change: true. Attributing everything to it: fake news.

[137] The El Niño signal began to drop in the late summer months of 2017, but it wasn't until the beginning of October that the Niño 3.4 zone temperature anomalies began to drop toward -1°C, which starts to signal La Niña and not a neutral ENSO state.

[138] Cal Fire, "Top 20 Most Destructive California Wildfires"
Jan 12, 2018
http://www.fire.ca.gov/communications/downloads/fact_sheets/Top20_Destruction.pdf

[139] National Weather Service San Francisco, Public Information Statement issued at 5:30 AM Oct 9, 2017
https://nwschat.weather.gov/p.php?pid=201710091230-KMTR-NOUS46-PNSMTR

[140] Wikipedia, Tubbs Fire
https://en.wikipedia.org/wiki/Tubbs_Fire

[141] Cal Fire, Incident Information, Jan 24, 2018
http://cdfdata.fire.ca.gov/incidents/incidents_stats?year=2017

[142] Wikipedia , 2017 California Wildfires retrieved 1/27/2018
https://en.wikipedia.org/wiki/2017_California_wildfires

[143] San Francisco Chronicle, "PG&E says someone else's wires may have started Tubbs Fire" David Baker and Evan Sernoffsky, Nov 9, 2017
http://www.sfchronicle.com/business/article/PG-E-says-someone-else-s-wires-may-have-started-12346129.php

[144] KRCR News, ABC 7, "Helena & Fork Fires Sept. 13 Update" by Haleigh Pike Sep 14, 2017
http://krcrtv.com/news/local/helena-fork-fires-sept-13-update-21124-acres-burned-53-percent-contained

[145] There were two Canyon fires. One was apparently human caused, and the second was started by embers from the first. Here's info on the first:
NBC LA Channel 4, "Fire Officials Reveal Causes of Orange County Canyon Fires" Nov 6, 2017
https://www.mercurynews.com/2017/10/10/pge-power-lines-linked-to-wine-country-fires/

[146] The Mercury News "PG&E power lines linked to Wine Country fires" Paul Rogers,

Oct 13, 2017
https://www.mercurynews.com/2017/10/10/pge-power-lines-linked-to-wine-country-fires/

[147] The 45 mph winds were recorded at Pt. Mugu 10/9/17, and 64-75 mph winds were recorded at Boney Mountain in Newbury Park, CA.

[148] Camarillo Airport reported 106° Oct 23, 2017 at 4:55 PM local time. A low temp that night dropped to 62°, making for a day-to-night difference of 42°. An all-time record of 108° was recorded at Camarillo Airport Oct 9, 2015. Data obtained using archived data from NOAA's National Centers for Environmental Information going back to 1950

[149] Using archived data from NOAA's National Centers for Environmental Information going back to 1950, a wind storm on November 21, 1957 blew winds to 57.5 mph from the northeast. Interestingly enough, daytime temps only reached 74°, which is a bit low for that type of Santa Ana. Other major wind events occurred in 1959, 1962, 1963, 1964, 1966, 1969 that exceeded 50 mph during a Santa Ana event.

[150] The Tribune, San Luis Obispo, "SoCal Edison workers started devastating Thomas Fire, victims' lawsuit says" Andrew Sheeler Jan 5, 2018
http://www.sanluisobispo.com/news/state/california/fires/article193262809.html

[151] VC Star, "Lawsuits allege Southern California Edison negligently started Thomas Fire" Mike Harris Jan 4, 2018
http://www.vcstar.com/story/news/local/2018/01/04/lawsuits-allege-southern-california-edison-negligently-started-thomas-fire/991192001/

[152] Zillow, "1048 Sunnycrest Ave Ventura, CA 93003" like other homes in that area show construction material to be "Composition Shingle"
https://www.zillow.com/homes/for_sale/16321142_zpid/34.302509,-119.196156,34.283505,-119.214717_rect/15_zm/1_fr/

[153] City of Ventura "Materials And Construction Methods For Exterior Wildfire Exposure". See 704A.3 item #3.
https://www.cityofventura.ca.gov/DocumentCenter/View/11291

[154] Comparing open-space backed homes to others, see 968 Scenic Way Dr, Ventura compared to 7159 Ridgecrest Ct across the street. The Scenic Way property is well over $1M while the Ridgecrest is $873K. Source Zillow.

[155] CNN "As the California wildfires consumed houses, five friends grabbed garden hoses and went to work" Doug Criss, Dec 6, 2017
https://www.cnn.com/2017/12/05/us/california-wildfire-fighting-with-hoses-trnd/index.html

[156] Los Angeles Times, "With more than 8,500 firefighters doing battle, this is California's largest wildfire response" Dec 18, 2017 Melissa Etehad and Brittny Mejia
http://www.latimes.com/local/lanow/la-me-thomas-fire-epic-fight-20171218-htmlstory.html

---

[157] U.S. Forest Service, The National Wildfire Coordinating Group, Thomas Fire incident information, retrieved 1/29/18 at:
https://inciweb.nwcg.gov/incident/5670/

[158] NASA has some incredible images taken from space of the Thomas and other fires. Check this out:
https://earthobservatory.nasa.gov/NaturalHazards/view.php?id=91379

[159] We saw the flames from our house in Newbury Park, 15 miles away from the fire. It was clear as could be at night, and it gave the optical illusion that the fire was coming right for us. I have some shots on Facebook, one of which is from the front of our house here:
https://www.facebook.com/photo.php?fbid=10214511197998134&set=a.1622459205792.2084317.1365836460&type=1&theater

[160] Those clear nights as the fire burned, visibility spanned to the horizon. But, you couldn't see land features like hills and such, and without them it gave an illusion that the fire was approaching your backyard. We watched this from our bedroom window many times during the fire, but we were 15 miles away in Thousand Oaks, CA.

[161] KSBY "Recovery Mode: More than 26k tons of Thomas Fire debris removed in Ventura County" Kayla Cash Feb 13, 2018
http://www.ksby.com/story/37495758/recovery-mode-more-than-26k-tons-of-thomas-fire-debris-removed-in-ventura-county

[162] Before and after pictures were a combination of Google Earth images compared to images shot by journalists. Here are some sources:
SF Gate, "Before-and-after photos show the devastating roll of Southern California wildfires" Michelle Robertson Dec 11,2017
https://www.sfgate.com/bayarea/article/Before-and-after-photos-Southern-California-fires-12422395.php
Business Insider "Before-and-after photos show how Ventura County has been ravaged by the wildfire" Jeremy Berke Dec 15, 2017
http://www.businessinsider.com/ventura-county-fire-before-and-after-photos-2017-12

[163] Los Angeles Daily News "Firefighter killed in Thomas fire 'died a hero,' chief says at his funeral" Dec 23, 2017
https://www.dailynews.com/2017/12/23/firefighter-killed-in-thomas-fire-died-a-hero-chief-says-at-his-funeral/

[164] Santa Barbara Independent, "Thomas Fire Fatality Identified" Kelsey Brugger Dec 13, 2017
https://www.independent.com/news/2017/dec/13/thomas-fire-fatality-identified/

[165] Lilac Fire killed at least 35 horses. The San Diego Tribune, "At least 35 horses died in the Lilac fire at San Luis Rey Downs training center" David Hernandez, Dec 8, 2017
http://www.sandiegouniontribune.com/news/wildfire/sd-me-fire-horses-20171208-story.html
And another approximately 30 horses died in the Creek Fire
Los Angeles Times, "Nearly 30 horses found burned to death by Creek fire in Sylmar"

Brittny Mejia, Dec 6, 2017
http://www.latimes.com/local/lanow/la-me-ln-horses-die-in-creek-fire-20171206-story.html

[166] Offshore winds became strong starting on Dec 4, 2017, and then continued with speeds in Camarillo at over 30 mph for a number of days, and over 20 mph for other days. Besides a brief break on Dec 13, 2017, Santa Ana winds blew continuously for 14 straight days, and then blew lightly off and on through the remainder of the December.

[167] IPCC AR5, The Physical Science Basis, Chapter 10, page 1459 section 26.4.2, "Tree Mortality and Forest Infestations"

[168] The Global and Mail, "Almost half of fires reported in B.C. this year caused by humans: province" Xioa Xu, July 10, 2017
https://www.theglobeandmail.com/news/british-columbia/almost-half-of-fires-burning-across-bc-caused-by-humans-province/article35652545/

[169] Popular Science "This is how much of the world is currently on fire" Kendra Pierre-Louis, Aug 4, 2017
https://www.popsci.com/global-wildfire-maps#page-13

[170] Based on 12,750 KM diameter of the earth, and then halving for the radius.

[171] IPCC AR5, Working Group II: Impacts, Adaptation, and Vulnerability
Chapter 4.5 Human Settlements, Energy, and Industry, Table TS-3
http://www.ipcc.ch/ipccreports/tar/wg2/index.php?idp=35

[172] Google Trend with the following comparisons: "climate change landslide", "climate change mudslide", "climate change heatwave", "climate change flood"
Retrieved 2/9/2018 at:
https://trends.google.com/trends/explore?date=2017-01-01%202017-12-0&q=climate%20change%20landslide,climate%20change%20sea%20level,climate%20change%20heat%20wave,climate%20change%20flood

[173] USGS "Fire-Induced Water-Repellent Soils: An Annotated Bibliography" 1997 Mary Kalendovsky and Susan Cannon
https://pubs.usgs.gov/of/1997/720/pdf/OFR-97-720.pdf

[174] National Weather Service, "How It All Came Down :Weather conditions leading to the Christmas Flood of 1964" by Dave Elson and Andy Bryant
http://www.nwp.usace.army.mil/Portals/24/docs/water/64_flood_wx.pdf

[175] USGS "Southern California Landslides — An Overview"
https://pubs.usgs.gov/fs/2005/3107/pdf/FS-3107.pdf

[176] Santa Barbara County Executive Office Press Release Jan 8, 2018
https://www.countyofsb.org/asset.c/3634

[177] KPCC "Santa Barbara County issued conflicting mudslide warnings" Jan 23, 2018
https://www.scpr.org/news/2018/01/23/80049/california-officials-lifting-some-mudslide-

evacuat/

[178] Wikipedia "2018 Southern California mudflows"
https://en.wikipedia.org/wiki/2018_Southern_California_mudflows#Santa_Barbara_County

[179] This and much of the information on the mishaps on alerts is from :the Los Angeles Times, Urgent alerts about deadly mudslides came too late for Montecito victims" Joseph Serna, Haily Branson-Potts, Ruben Vives, Laura Nelson, Jan 11, 2018
http://www.latimes.com/local/lanow/la-me-mudslide-factors-20180111-story.html

[180] The story of Josie Gower and her boyfriend Norm were referenced from two stories: The Los Angeles Times, "Urgent alerts about deadly mudslides came too late for Montecito victims" Joseph Serna, Haily Branson-Potts, Ruben Vives, Laura Nelson, Jan 11, 2018
http://www.latimes.com/local/lanow/la-me-mudslide-factors-20180111-story.html
and
ABC 7 Eyewitness News, "Names of Montecito mudslide victims released" Jan 20, 2018
http://abc7.com/names-of-montecito-victims-released/2930517/

[181] Santa Barbara County Sheriff's Office "Search Teams Locate Body of Missing Montecito Mother" Jan 21, 2018
https://www.sbsheriff.org/search-teams-locate-body-missing-montecito-mother/

[182] Twitter, SB Shariff's Office Jan 31, 2018
https://twitter.com/sbsheriff/status/958964329151217664

[183] The Tribune, "Mudslides took out Montecito's water supply. SLO County crews hiked 14 miles to fix it" Monica Vaughan Feb 2, 2018
http://www.sanluisobispo.com/news/state/california/article198206534.html

[184] KEYT "Where is the mud and debris from Montecito going?" Oscar Flores Jan 12, 2018
http://www.keyt.com/news/environment/where-is-the-mud-and-debris-from-montecito-going/685208063

[185] Both SBCPH and VCRMA had weekly testing results that continued to show excessive bacteria counts on Jan 31, 2018.
News of continued dumping was found in the Santa Barbara Independent, "Detective Work of Montecito Mud at Goleta Beach" Melinda Burns, Feb 1, 2018 at:
https://www.independent.com/news/2018/feb/01/detective-work-montecito-mud-goleta-beach/

[186] SBCPHD press release regarding public health advisory
http://cosb.countyofsb.org/uploadedFiles/phd/Press_Release/2018_Press_Releases/2018-01-17%20Health%20Advisory%20PR.pdf

[187] KEYT, "Managing the Montecito disaster mud has homeowners looking at options" John Palminteri Feb 13, 2018
http://www.keyt.com/news/santa-barbara-s-county/managing-the-montecito-disaster-

mud-has-homeowners-looking-at-options/701178828

[188] K PCC "Santa Barbara County issued conflicting mudslide warnings" Jan 23, 2018 https://www.scpr.org/news/2018/01/23/80049/california-officials-lifting-some-mudslide-evacuat/

[189] The Weather Channel "Man Swept Away Off Oregon Coast Presumed Dead, Authorities Say" January 19, 2018
https://weather.com/news/news/2018-01-18-oregon-coast-strong-waves
and
The Oregonian "Police identify man swept into ocean in Depoe Bay" Anna Marum Jan 19, 2018
http://www.oregonlive.com/pacific-northwest-news/index.ssf/2018/01/police_identify_man_swept_into.html
and
Newport News Times "Man swept out to sea identified, still lost" Feb 3, 2018 Rick Beasley
https://newportnewstimes.com/article/man-swept-out-to-sea-identified-still-lost

[190] Los Angeles Times "Rain moves in to Southern California, prompting closure of Highway 33 near Ojai" Mar 2, 2018 Melissa Etehad and Hailey Branson-Potts
http://www.latimes.com/local/lanow/la-me-ln-winter-storm-20180302-story.html

[191] CNN "Landslide buries California's scenic highway in Big Sur" Nicole Chavez, Jason Hanna, Keith Allen May 25, 2017
http://www.cnn.com/2017/05/24/us/california-landslide-scenic-highway/index.html

[192] The Tribune "An exclusive look at the new Hwy. 1 over Big Sur's massive landslide" Stephen H. Provost Jan 8, 2018
http://www.sanluisobispo.com/news/local/article193534584.html

[193] Wikipedia 1995 La Conchita landslides
https://en.wikipedia.org/wiki/2005_La_Conchita_landslide#1995_La_Conchita_landslides

[194] Storm details were derived from NOAA's NCDC archive database at:
https://www.ncdc.noaa.gov/stormevents/eventdetails.jsp?id=5439118

[195] Zillow, 6991 Vista Del Rincon Dr, 4 bedrooms, 2 baths, 1,828 square feet accessed 2/2/18 from
https://www.zillow.com/homes/for_sale/16318733_zpid/globalrelevanceex_sort/34.379093,-119.434583,34.343844,-119.468658_rect/14_zm/

[196] Here's a great example where 7170 Carpentaria Ave, a 2 bed 2 bath 1,400 square foot home sole in February 2015 for $423,500 but the Zestimate was almost $1.7 million
https://www.zillow.com/homes/for_sale/16318760_zpid/globalrelevanceex_sort/34.379093,-119.43188,34.343844,-119.471405_rect/14_zm/

[197] Los Angeles Times, "Flames reach small seaside town of La Conchita, sending evacuation holdouts fleeing" Ruben Vives, Alene Tchekmedyian Dec 7, 2017

http://www.latimes.com/local/lanow/la-me-ln-la-conchita-thomas-fire-20171207-story.html

[198] Wikipedia, 2017 Sierra Leone Mudslides
https://en.wikipedia.org/wiki/2017_Sierra_Leone_mudslides

[199] BBC News "Freetown: A disaster waiting to happen?" Flora Drury Aug 16, 2017
https://www.bbc.com/news/amp/world-africa-40933491

[200] The World Bank, "Sierra Leone, rapid damage and loss assessment of August 14th, 2017 landslides and floods in the western area"
https://reliefweb.int/sites/reliefweb.int/files/resources/19371_Sierra_Leone_DaLA_Web-forprinting.pdf

[201] Director Lewin's quote from the Los Angeles Times, "Urgent alerts about deadly mudslides came too late for Montecito victims" Joseph Serna, Haily Branson-Potts, Ruben Vives, Laura Nelson, Jan 11, 2018
http://www.latimes.com/local/lanow/la-me-mudslide-factors-20180111-story.html

[202] Wikipedia "Effect of Hurricane Katrina on the Louisiana Superdome"
https://en.wikipedia.org/wiki/Effect_of_Hurricane_Katrina_on_the_Louisiana_Superdome

[203] This is about the only last, publicly available mention of the old WRIP project from NOAA:
http://www.nws.noaa.gov/om/notification/pns12wrip_ote.htm

[204] After hearing an Uber pitch in December 2008, Uber got funding in March 2009. See: Business Insider, "The story of how Travis Kalanick built Uber into the most feared and valuable startup in the world" Avery Hartmans and Nathan McAlone Aug 1, 2016
http://www.businessinsider.com/ubers-history/#june-1998-scour-a-peer-to-peer-search-engine-startup-that-kalanick-had-dropped-out-of-ucla-to-join-snags-its-first-investment-from-former-disney-president-michael-ovitz-and-ron-burkle-of-yucaipa-companies-1

[205] One notorious snoozer was in charge of the NOAA Weather Wire Service (NWWS), Martin Baron, or "Marty". A nice, quiet man, but no one really seemed to know what he did. I don't recall who all asked to be added to the meeting minutes, but I don't believe Marty did; in fact, he was either sleeping or absent from most meetings.

[206] While I can't divulge all the details of the languages and systems, I can say that the language of choice by the contractor was one supplanted in the 1990s.

[207] Washington Post, "National Weather Service budget cuts misguided, misplaced" Steve Tracton, March 29, 2012
https://www.washingtonpost.com/blogs/capital-weather-gang/post/national-weather-service-budget-cuts-misguided-misplaced/2012/03/29/gIQAmm6qiS_blog.html?utm_term=.45de7a61164e

[208] Burlington Free Press, "Weather alert: Forecaster shortage risks public safety, NWS

workers say" Adam Silverman Oct 20, 2017
http://www.burlingtonfreepress.com/story/news/2017/10/20/national-weather-service-forecaster-shortage-risks-public-safety-nws-workers-say/755653001/

[209] The Washington Post "White House proposes steep budget cut to leading climate science agency" Steven Mufson, Jason Samenow, Brady Dennis Mar 3, 2017
https://www.washingtonpost.com/news/energy-environment/wp/2017/03/03/white-house-proposes-steep-budget-cut-to-leading-climate-science-agency/?utm_term=.23faeb81430c

[210] NWS, "History of the National Weather Service"
https://www.weather.gov/timeline

[211] FCC, Wireless Emergency Alerts
https://www.fcc.gov/consumers/guides/wireless-emergency-alerts-wea

[212] CNN "Here's what went wrong with the Hawaii false alarm" Madison Park, Jan 31, 2018
https://www.cnn.com/2018/01/31/us/hawaii-false-alarm-investigation-findings/index.html

[213] Santa Barbara Independent, "When Evacuation Orders Go Wrong" Keith Hamm Dec 13, 2017
https://www.independent.com/news/2017/dec/13/when-evacuation-orders-go-wrong/

[214] iPhone instructions for receiving WEA alerts:
https://www.verizonwireless.com/support/knowledge-base-206948/
For all devices from Verizon, see:
https://www.verizonwireless.com/support/wireless-emergency-alerts-compatible-devices/
Or contact your carrier for more information.

[215] County of Santa Barbara, "72-Hour Storm Evacuation Timeline"
http://countyofsb.org/asset.c/3922

[216] Without Googling, it, can you tell the difference between "Elevated" and "Guarded" warnings from the Department of Homeland Security? OK, now you can see the answer at:
https://en.wikipedia.org/wiki/Homeland_Security_Advisory_System

[217] SF Gate, "Feinstein, Harris criticize federal alert system as inadequate" Joaquin Palomino Oct 17, 2017
http://www.sfgate.com/news/article/Feinstein-Harris-criticize-federal-alert-system-12286268.php
Also, on death toll in NorCal, see Eyewitness News ABC7 "Thomas Fire: State legislators call for emergency alert system in response to massive blaze" Jan 4, 2018 at:
http://abc7.com/thomas-fire-state-legislators-call-for-emergency-alert-system/2862395/

[218] FEMA Fact Sheet, "Alerting Authority Guidance for Issuing Alert Messages"
https://www.fema.gov/media-library-data/1409250796582-

6a43b7939639ae6bdf565cf35cece90b/WEA%20ALERTING%20CRITERIA%20fact%20sheet%208-4-2014_508.pdf

[219] FEMA "Frequently Asked Questions: Wireless Emergency Alerts" retrieved 2/2/2018, "What does a WEA message look like? …the message will be no more than 90 characters"
https://www.fema.gov/frequently-asked-questions-wireless-emergency-alerts

[220] CNN "Trump's pick to lead NOAA pushed for privatizing weather data" Miranda Green Oct 14, 2017
https://www.cnn.com/2017/10/14/politics/noaa-nominee-accuweather/index.html

[221] OpenSecrets.org Accuweather Inc retrieved 2/2/2018 at:
https://www.opensecrets.org/lobby/clientsum.php?id=F106514&year=2017

[222] Becker's Hospital Review, "25 things to know about Blue Cross Blue Shield" Emily Rappleye June 10, 2015
https://www.beckershospitalreview.com/payer-issues/25-things-to-know-about-blue-cross-blue-shield.html

[223] The Weather Channel, "2017 Off to Destructive Start: Severe Weather Reports Tally 5,000+, More Than Double the Average" Chris Dolce Apr 10, 2017
https://weather.com/storms/tornado/news/severe-weather-hail-tornado-wind-damage-2017-mid-april

[224] You can see more at the USGCRP website at:
https://www.globalchange.gov/

[225] CSSR Executive Summary can be found at:
https://science2017.globalchange.gov/chapter/executive-summary/

[226] California Air Resources Board, "The Governor's Climate Change Pillars: 2030 Greenhouse Gas Reduction Goals", retrieved 3/1/2018
https://www.arb.ca.gov/cc/pillars/pillars.htm

[227] California Climate Change website, "California Climate Change Executive Orders" Retrieved 3/1/2018
http://www.climatechange.ca.gov/state/executive_orders.html

[228] The Mercury News "Oroville Dam: Feds and state officials ignored warnings 12 years ago" Feb 12, 2017 Paul Rogers
https://www.mercurynews.com/2017/02/12/oroville-dam-feds-and-state-officials-ignored-warnings-12-years-ago/

[229] The Sacramento Bee, "California Legislature votes to keep dam-safety plans secret" June 15, 2017 Ryan Sabalow, Jim Miller, Dale Kasler
http://www.sacbee.com/news/state/california/water-and-drought/article156379919.html

[230] The Climate Reality Project
https://www.climaterealityproject.org/our-mission

---

[231] Charity Navigator, "The Climate Reality Project" retrieved 3/1/2018 https://www.charitynavigator.org/index.cfm?bay=search.summary&orgid=12771
[232] The Huffington Post, "Al Gore's Stupendous Wealth Complicates His Climate Message. That Can Change." Alexander C. Kaufman Aug 18, 2017 https://www.huffingtonpost.com/entry/al-gore-wealth_us_599709f2e4b0e8cc855d5c09

Made in the USA
San Bernardino, CA
28 March 2019